ELECTRICAL INSTALLATION COMPETENCES

Part 2 Studies: Science

Maurice Lewis
BEd (Hons), FIEIE

Stanley Thornes (Publishers) Ltd

First published in 1993 by:
Stanley Thornes (Publishers) Ltd
Ellenborough House
Wellington Street
CHELTENHAM
GL50 1YD
UK

A catalogue record for this book is available from the British Library.
ISBN 0 7487 1660 2

Typeset by Florencetype, Kewstoke, Avon
Printed and bound in Great Britain by Scotprint Ltd, Musselburgh

Contents

Preface

This book is written for Part 2 students studying the City and Guilds of London Institute Course 236–8 syllabus in electrical installation competences. It is the second book in a series of three books aimed at covering electrical science topics of the current syllabus. To help you cope with the technicalities of the course, it was decided to include a revision of *basic mathematics*. The first chapter, therefore, covers algebraic equations, use of a scientific calculator, transposition of formulae, simultaneous equations and solving problems using trigonometrical ratios.

Chapter 2 concerns *alternating current circuits*, introducing you to phase displacement between current/voltage quantities when applied to circuit components possessing resistance, inductance and capacitance. You will be shown numerous worked examples of RLC components connected in series and parallel and you will also see how easy it is to construct graphs and phasor diagrams of circuits which illustrate power factor. The chapter introduces three-phase systems and balancing of loads, star and delta connections as well as providing a number of calculations on three-phase power involving power triangles using the units of kW, kVA and kVAr.

Chapter 3 deals with *motors and starters* describing the basic principles of motor operation of single-phase and three-phase motors. Starting methods are also discussed along with the protective measures to safeguard against undervoltage and overload. The chapter is supported with numerous, easy to draw, circuit diagrams for the reader.

Chapter 4 covers elements of *lighting design* and summarizes some of the common lamps in use today which operate from incandescent, fluorescent and discharge sources. This chapter supplements the studies of lighting found in Part 1: Theory and prepares student readers for studies at Part 3 level (i.e. the Course 'C' Certificate).

Each chapter includes appropriate exercises. In addition, Appendix 1 provides the student with 70 multiple choice questions; Appendix 2 sets 10 written questions; Appendix 3 provides answers to all exercises, multiple choice questions and written questions.

Maurice Lewis

Introductory mathematics

Objectives

After reading this chapter you should be able to:

- *construct equations from derived data.*
- *perform simple algebraic factorisation.*
- *multiply expressions in brackets.*
- *use standard form notation.*
- *solve simultaneous equations.*
- *rearrange formulae to change the subject.*
- *solve problems with the aid of a scientific calculator.*
- *solve triangles using sine, cosine and tangent ratios.*

Algebraic equations

The branch of mathematics in which letters and symbols are used to represent quantities is called **algebra**. One of its advantages is the freedom it allows to express unknown quantities when using different types of equation and formula. Another is its use in solving problems without the distraction created by difficult arithmetic. The following examples serve to illustrate these two points.

Create a simple algebraic equation for finding the distance round a room given that the distance (d) is equal to twice its length (l) and twice its breadth (b). Here:

$$d = 2l + 2b$$

or

$$d = 2(l + b)$$

To find the room's area (A) you would simply write:

$$A = l \times b$$

and to find its volume (V) you would write:

$$V = l \times b \times h$$

(where h is the height of the room). If numerical values replace the letter dimensions, then the answer for area and volume would need to be expressed in square units and cubic units respectively.

Create an equation for finding the total area of a rectangle if it is divided into two parts as shown in Figure 1.1. Here, the length of the two parts is called a and b and the width is called x. The areas of the two rectangles are, therefore, ax and bx respectively. The total area (A) is now expressed as:

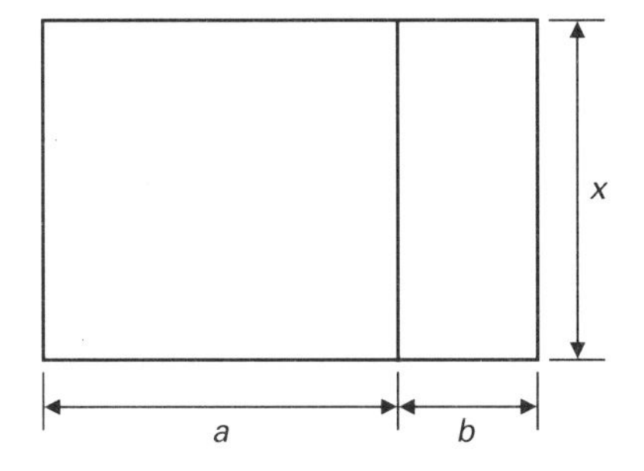

Figure 1.1 Area calculation

$$A = ax + bx$$

You saw in the first example how the equation was further simplified by inserting brackets. In this second example, you can see that x is common in both terms and brackets can be used to enclose the term $a + b$, simplifying the equation into:

$$A = x(a + b)$$

This process is called **factorisation** which is dealt with after the next example.

A 3 m length of conduit is to be cut into three pieces. If the second piece is twice as long as the first piece and the third piece is 40 cm shorter than the first piece, what are the lengths of all three pieces if 20 cm is allowed for wastage?

In this example, you must take the initiative of giving the first piece a letter symbol (say x), then the second piece becomes $2x$ and the third piece ($x - 40$). You should make a sketch of the problem (see Figure 1.2) and note that the useful length of conduit is only 280 cm. The problem is solved by creating a simple equation based on finding x. Hence

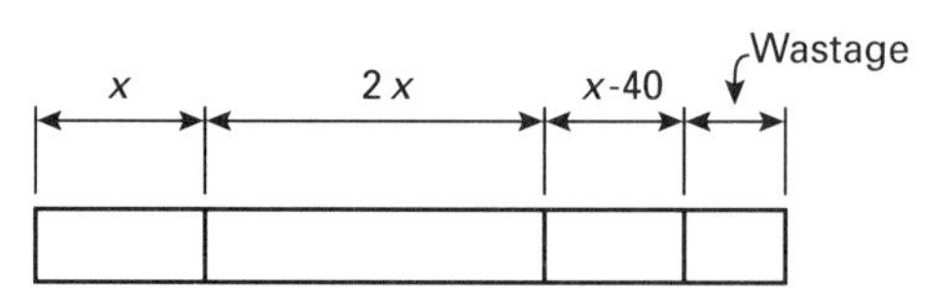

Figure 1.2 Length calculation

$$x + 2x + (x - 40) = 280$$
$$4x - 40 = 280$$
$$4x = 320$$

therefore $x = 80$ cm

hence, second piece

$$2x = 160 \text{ cm}$$

and third piece

$$x - 40 = 40 \text{ cm}$$

In this example, brackets are shown around the term $(x - 40)$ since it represents a collective term in itself but no factorising is required.

Factorisation

Factorise the following:

$$5x + 5y$$

Since 5 is common to both sides then $5(x + y)$.

$$7x^2 - 14x$$

Since 7x is common to both sides then $7x(x - 2)$.

$$2x + 120$$

Since 2 is common to both sides then $2(x + 60)$.

Removing brackets

Remove the brackets from the following:

$$3(x + y) = 3x + 3y$$
$$-4(x + y) = -4x - 4y$$
$$x(x - y) = x^2 - xy$$
$$4a(2a + b) = 8a^2 + 4ab$$

The laws of algebra are the same as those for arithmetic. In multiplication the sign is often omitted when writing down more than one term, e.g. $4xyz$ means $(4 \times x \times y \times z)$ but the sign must reappear when numbers are substituted for actual symbols. It is usual practice to assume that all symbols, letters and numbers have a plus value unless a minus sign is shown. You must remember the **sign rule** used in ordinary multiplication:

Rule: *like signs give a plus and unlike signs give a minus.*

Substitution

If x = 4 and y = 5, then:

$$3(x + y) = 3(4 + 5) = 27$$
$$-4(x + y) = -4(4 + 5) = -36$$
$$x(x - y) = 4(4 - 5) = -4$$

In the fourth example, if a = −2 and b = 4 then:

$$4a(2a + b) = 4(-2)\,[2\,(-2) + 4]$$
$$= -8\,(-4 + 4) = 0$$

Notice the different types of bracket used. The terms contained within the inner brackets must be tackled first before dealing with the group terms, separated in this example by the plus sign. See if you can solve the next problem.

Simplify the following algebraic expression by removing the brackets and then finding its value when $t = -1$

$$2(1 + 2t - 3t^3) - (3 - 4t + 5t^3)$$

The solution is:

$$-1 + 8t - 11t^3$$

and when $t = -1$ the numerical answer is 2.

Further examples

1) $$3(x - 13) = 9 - 3(x + 2)$$

Remove brackets:

$$3x - 39 = 9 - 3x - 6$$

collect like terms:

$$3x + 3x = 9 + 39 - 6$$

solve for x:

$$6x = 42$$
$$x = 7$$

2) $2(x + 1) = 5x - 7$

Remove brackets:

$$2x + 2 = 5x - 7$$

collect like terms:

$$2x - 5x = -7 - 2$$

solve for x:

$$-3x = -9$$

$$x = 3$$

3) $\frac{x - 2}{5} = x - 4$

Transpose 5:

$$x - 2 = 5(x - 4)$$

remove brackets:

$$x - 2 = 5x - 20$$

collect like terms:

$$x - 5x = 2 - 20$$

solve for x:

$$-4x = -18$$

$$x = 4.5$$

4) $(x + 3)(x + 5) = 55 + x^2$

Remove brackets:

$$x^2 + 3x + 5x + 15 = 55 + x^2$$

$$x^2 + 8x + 15 = 55 + x^2$$

$$x^2 - x^2 + 8x = 55 - 15$$

$$8x = 40$$

$$x = 5$$

Index notation

If a is any quantity and n is a positive integer (i.e. whole number) then a^n means $a \times a \times a \ldots$ to n factors. This is termed the index or nth power of a. The index thus indicates the number of times which a occurs as a factor.

Multiplication of powers

$$a^m \times a^n = a^{m+n}$$

If $a = 10$, m = 5 and n = 3 this becomes:

$$= 10^{5+3}$$

$$= 10^8$$

$$= 100\,000\,000$$

Further examples

$$5^2 \times 5^5 = (5 \times 5) \times (5 \times 5 \times 5 \times 5 \times 5)$$
$$= 5^{2+5}$$
$$= 5^7$$
$$8^3 \times 8^0 \times 8^4 = 8^{3+0+4}$$
$$= 8^7$$

Rule: *When multiplying powers of the same base, add the indices.*

Division of powers

$$a^m \div a^n = a^{m-n}$$

If we give m and n the same values as above, (i.e. $m = 5$ and n = 3) we have:

$$= 10^{5-3}$$

$$= 10^2$$

$$= 100$$

Note: If in the above example, $m = 3$ and n = 5, the answer would be 10^{-2} which is 1/100 or 0.01. You can show this by cancelling the number of tens in the numerator and denominator, for example:

$$\frac{10^m}{10^n} = \frac{\cancel{10} \times \cancel{10} \times 10}{\cancel{10} \times \cancel{10} \times \cancel{10} \times 10 \times 10}$$

$$= \frac{1}{100}$$

The line dividing 1 and 100 is called the **quotient line** and you will see that m (the **numerator**) is less than n (the **denominator**). This tells you that an index having a negative integer produces an answer as a fraction less than unity (e.g. 10^{-2} is really $1/10^2 = 0.01$).

You can see from the following list where the change from a positive integer to a negative integer occurs:

$$10^6 = 1\,000\,000$$
$$10^5 = 100\,000$$
$$10^4 = 10\,000$$
$$10^3 = 1000$$
$$10^2 = 100$$
$$10^1 = 10$$
$$10^0 = 1$$
$$10^{-1} = 1/10 = 0.1$$
$$10^{-2} = 1/100 = 0.01$$
$$10^{-3} = 1/1\,000 = 0.001$$
$$10^{-4} = 1/10\,000 = 0.0001$$
$$10^{-5} = 1/100\,000 = 0.000\,01$$
$$10^{-6} = 1/1\,000\,000 = 0.000\,001$$

Numbers in standard form

One way of avoiding errors with long complicated numbers, especially decimal numbers is to express them in standard form. This basically means multiplying them by a power of 10. You saw in the table above that $10^2 = 100$, $10^3 = 1\,000$ and $10^{-5} = 0.000\,01$. When the power of 10 is positive the numerical value of the index gives the number of places that the decimal point has to be moved to the right. If it is a negative value, the decimal point has to be moved to the left. Consider the following examples:

$$80\,987 = 80.987 \times 10^3$$

$$40\,000\,000 = 40 \times 10^6$$

$$259.885\,23 = 0.26 \times 10^3$$

$$0.094 = 9.4 \times 10^{-2}$$

$$0.\,000\,0345 = 3.45 \times 10^{-5}$$

$$2.463 \times 10^{-4} = 0.\,000\,246\,3$$

You can prove that $10^0 = 1$ by considering the expression:

$$a^0 = a^{n-n}$$
$$= a^n \div a^n$$
$$= \frac{a^n}{a^n} = 1$$

Note also

$$a^0 \times a^n = a^{0+n}$$
$$= a^n$$

Therefore

$$a^0 = 1$$

Further examples

$$\frac{4^9}{4^3} = 4^{9-3} = 4^6$$

$$\frac{a^4}{a^4} = a^{4-4} = a^0 = 1$$

Rules: *When dividing powers of the same base subtract the index of the denominator from the index of the numerator. Any base raised to the index of zero is equal to 1.*

Powers of powers and fractional powers

Indices can take the form of a power of a power, i.e. $(a^m)^n$ and also be a fractional power, i.e. $a^{1/2}$. In the former case, the two powers are multiplied together whereas in the latter case the power is more conveniently written as the square root of a, i.e. $\sqrt{a}$.

Note: $a^{1/2} \times a^{1/2} = a^1 = a$ (e.g. $\sqrt{3} \times \sqrt{3} = 3$).

Further examples

$$(10^6)^3 = 10^{6\times3} = 10^{18}$$

$$\left(\frac{m^2}{n^3}\right)^3 = \frac{(m^2)^3}{(n^3)^3} = \frac{m^{2\times3}}{n^{3\times3}} = \frac{m^6}{n^9}$$

$$a^{-n} = \frac{1}{n}$$

Rules: *When raising the power of a base to a power, multiply the indices together. The power of a base which has a negative index is the reciprocal of the power of the base with the same but positive index.*

Exercise 1.1

1. Simplify $a^3b + ab^2$
2. Simplify $a^3b \times ab^2$
3. Figure 1.3 shows a triangle ABC which is divided into two right-angled triangles by the perpendicular line h. Derive an equation for finding its area.

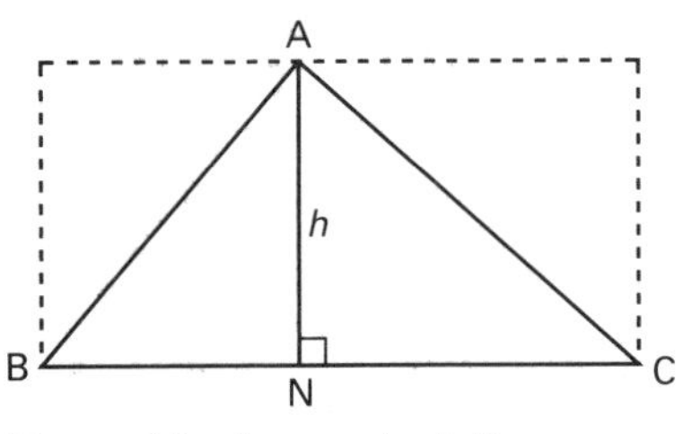

Figure 1.3 Area calculation

Hint: Treat both small triangles as half rectangles and note that the line BC = BN + NC

4. Figure 1.4 shows a rectangle with two triangles cut out. Write down an algebraic equation for the area of the remainder and solve the equation when l = 15 cm, b = 10 cm and h = 4 cm

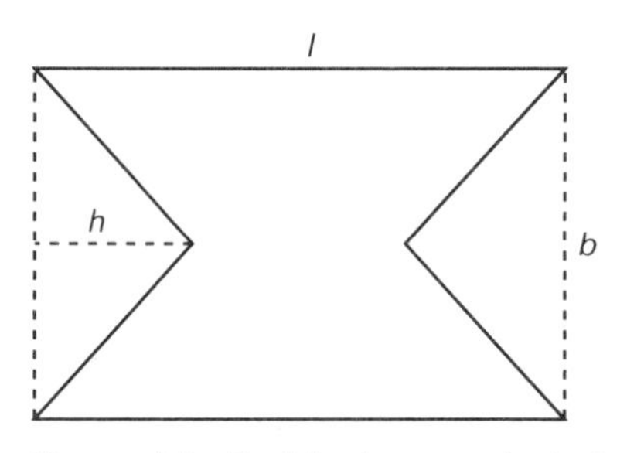

Figure 1.4 Residual area calculation

5. Solve the equation $2(x + 5) = 3(2x - 4)$
6. Two rectangular boards are equal in cross sectional area. The length of one is 18 cm and that of the other 16 cm. If the difference in their breadths is 4 cm, find the breadth of each board and their common area.
7. Create a simple algebraic equation for finding the area (wall thickness) of a standard piece of metal conduit.
8. The sum of resistance for three resistors connected in series with each is given by

$$\Sigma R = R_1 + R_2 + R_3.$$

If $R = V/I$, create a formula for potential difference (V).

9. Figure 1.5 shows a radial distributor AG. If the current flowing into point A is x amperes, state with reference to x the current flowing between each section and also the current flowing towards G.
10. Figure 1.6 shows a right-angled triangle representing power quantities viz. power (P), voltamperes (S) and reactive voltamperes (X). Write down an expression for finding X in terms of the other two quantities using Pythagoras' theorem.

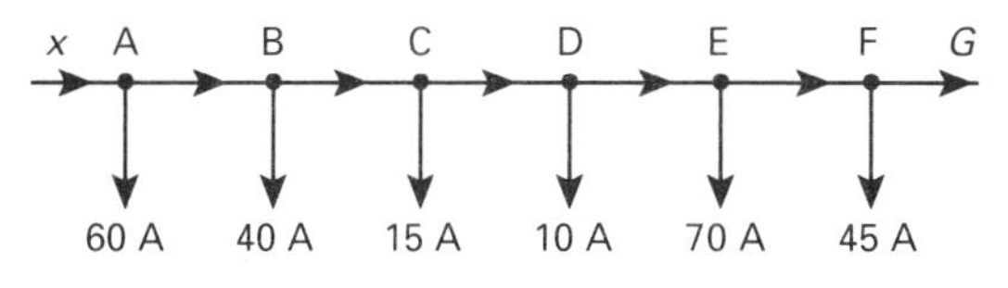

Figure 1.5 Radial distributor

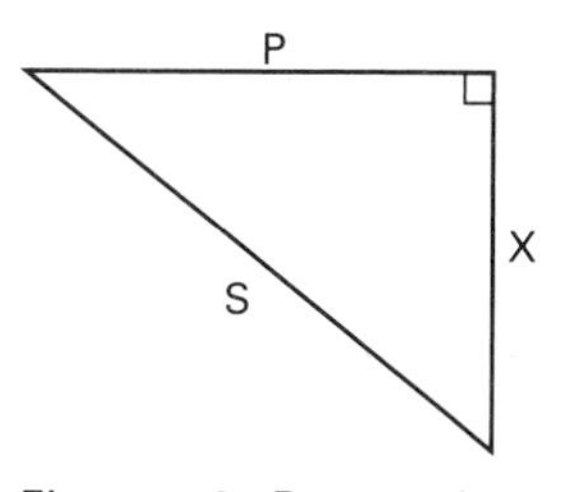

Figure 1.6 Power triangle

Simultaneous equations

These are algebraic equations containing two (or more) unknown quantities such as x and y. The equation is solved by using information or data extracted from the problem. The following examples illustrate the method used.

Example 1.1

To find two numbers x and y if their sum is 100 and their difference is 20 is expressed as:

$$x + y = 100 \qquad [1]$$

and

$$x - y = 20 \qquad [2]$$

Solution

To eliminate y add together both equations.

Thus

$$2x = 120$$

and therefore

$$x = 60$$

If x is substituted in equation [1] you can solve for y:

Hence

$$60 + y = 100$$

and therefore

$$y = 100 - 60$$
$$= 40$$

Example 1.2

An electrical contractor installed in a premises x luminaires which cost £30 each and y electric fires which cost £50 each. If the total expenditure was £490 and he spent £10 more on the fires than he did on the luminaires, determine how many luminaires and fires he purchased.

Solution

Here, the two equations needed to solve the problem are:

$$30x + 50y = 490 \qquad [1]$$

$$-30x + 50y = 10 \qquad [2]$$

Equation [2] tells you that the total cost of y is more than the total cost of x by £10.

Adding the equations together will result in:

$$100y = 500$$

and therefore

$$y = 5$$

Substituting $y = 5$ in equation [1] solves for x:

Hence

$$30x + 250 = 490$$

$$30x = 490 - 250$$

$$30x = 240$$

and therefore

$$x = 8$$

To solve simultaneous equations it is often necessary to multiply or divide the equations to make the coefficients of one of the unknowns the same in both equations. It is then a matter of adding or subtracting the two equations to eliminate one of the unknowns. The next three problems illustrate this method.

Example 1.3

From the information given, solve for x and y:

$$3x - 5y = 6 \qquad [1]$$

$$4x + 3y = 37 \qquad [2]$$

Solution

Multiply equation [1] by 3 and equation [2] by 5. This will make the y terms the same in both equations and since they have opposite signs their sum will be zero.

Hence

$$9x - 15y = 18 \qquad [3]$$

and

$$20x + 15y = 185 \qquad [4]$$

By adding both these equations:

$$29x = 203$$

then

$$x = \frac{203}{29} = 7$$

Substituting this value in equation (1) gives:

$$21 - 5y = 6$$

$$-5y = 6 - 21$$

$$-5y = -15$$

therefore $\quad y = 3$

You should check your answer by substituting x and y in the second equation, to give $(4 \times 7) + (3 \times 3) = 37$

Example 1.4

An electrical contracting firm pays its senior apprentices £x per hour for their basic pay and £y per hour for their overtime. One apprentice works a basic week of 38 hours plus 6 hours overtime while a second apprentice works a basic week of 40 hours plus 4 hours overtime. If the first apprentice is paid £235 and the second apprentice is paid £230, determine their basic and overtime pay per hour.

Solution

In this example the equations are:

$$38x + 6y = 235 \quad [1]$$

and

$$40x + 4y = 230 \quad [2]$$

Multiply equation [1] by 2 and equation [2] by 3 and then subtract equation [1] from equation [2] in order to eliminate y. The rule for subtracting is to change all the signs in the equation you are concerned with, then add it to the second equation.

Hence

$$76x + 12y = 470 \quad [3]$$

and

$$120x + 12y = 690 \quad [4]$$

Now change signs of equation [3] and add to equation [4]:

$$-76x - 12y = -470 \quad [3']$$

$$120x + 12y = 690 \quad [4]$$

$$44x = 220$$

therefore

$$x = 5$$

Substitute x in equation (1)

Hence

$$190 + 6y = 235$$

$$6y = 235 - 190$$

therefore

$$y = 7.5$$

The basic rate of pay is therefore £5/hour and the overtime pay is £7.50/hour.

Example 1.5

Figure 1.7 shows a closed loop circuit in which two batteries send current to different parts of the circuit. Derive equations for the current flowing round loop ABEF and loop ACDF and solve the equation for the unknown currents. Assume $I_1 + I_2 = I_3$.

Solution

Since the batteries are the driving force, the

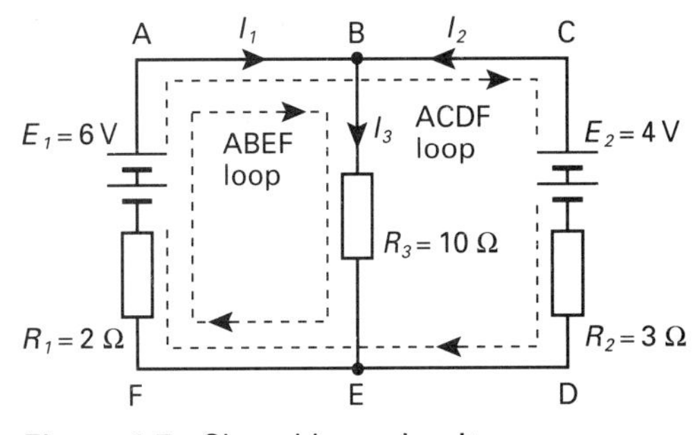

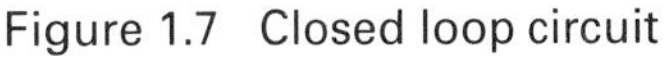

Figure 1.7 Closed loop circuit

currents to each loop can be expressed as:

$$I_1 R_1 + (I_1 + I_2)R_3 = E_1 \quad [1]$$

and

$$I_1R_1 - I_2R_2 = E_1 - E_2 \quad [2]$$

Inserting the values in these two equations gives:

$$2I_1 + 10\,(I_1 + I_2) = 6 \quad [1]$$

and

$$2I_1 - 3I_2 = 2 \quad [2]$$

The two equations become:

$$12I_1 + 10I_2 = 6 \quad [1]$$

and

$$2I_1 - 3I_2 = 2 \quad [2]$$

Multiply equation [2] by 6 and subtract it from equation [1] which will give the answer to I_2:

$$12I_1 + 10I_2 = 6 \quad [3]$$

and

$$12I_1 - 18I_2 = 12 \quad [4]$$

Subtracting equation [4] gives:

$$28I_2 = -6$$

therefore

$$I_2 = -0.214 \text{ A}$$

The minus sign shows that I_2 is flowing against I_1. If this value is now substituted in equation [2], I_1 can be found:

Thus

$$2I_1 - 0.\,642 = 2$$

and

$$2I_1 = 1.358$$

therefore

$$I_1 = 0.679 \text{ A}$$

The third branch current is

$$I_3 = I_1 + I_2 = 0.679 - 0.214$$
$$= 0.465 \text{ A.}$$

Q

Example 1.6

Sometimes the information extracted from a problem results in three related equations to be solved. Consider the following but remember to change the signs of an equation when subtracting it and adding it to another equation:

Solution

$$x + y + z = 53 \quad [1]$$

$$x + 2y + 3z = 105 \quad [2]$$

$$x + 3y + 4z = 134 \quad [3]$$

Subtracting equation [1] from equation [2] gives:

$$y + 2z = 52 \quad [4]$$

and subtracting equation [2] from equation [3] gives:

$$y + z = 29 \quad [5]$$

Now see if you can finish off this problem along the lines explained above.

Hints:

Subtract equation [5] from equation [4] to give $z = 23$

Substitute z in equation [5] to give $y = 6$

Substitute y and z in equation [1] to give $x = 24$.

Transposition of formulae

This topic was introduced in the Part 1 Science book and is reviewed here in more detail since it still presents a problem to most Part 2 students. Transposition of formulae requires not only an understanding of the basic rules of arithmetic but also the ability to manipulate terms and symbols in order to change their position or place. The intention is to nominate a new quantity to become the subject of the formula.

The first rule you must learn is that everything on either side of an equal sign, i.e. the left hand side (LHS) and right hand side (RHS) is in a state of balance. You cannot do anything to one side without it affecting the other side. The following example illustrates this point using numbers instead of letters to make the procedure easily understood.

$$2 \times 5 = 1 \times 10$$

If you *add* a number (say 60) to the LHS of the equal sign, you will have to repeat this on the RHS. You must think of the equal sign as a pivot balancing the two sides like a set of scales:

$$60 + (2 \times 5) = 60 + (1 \times 10)$$

Note: The brackets must be used to keep the original numbers together.

If you *subtract* (say 40) from the LHS, you must repeat this for the RHS:

$$\frac{60 - 40 + (2 \times 5)}{2} = \frac{60 - 40 + (1 \times 10)}{2}$$

If you *divide* by 2 on the LHS, you must repeat this on the RHS:

$$\frac{60 - 40 + (2 \times 5)}{2} = \frac{60 - 40 + (1 \times 10)}{2}$$

And if you *multiply* by 4, the LHS and RHS both become:

$$\frac{4[60 - 40 + (2 \times 5)]}{2} = \frac{4[60 - 40 + (1 \times 10)]}{2}$$
$$= 60$$

You should take note of the way the extra brackets have been inserted so that the number 4 embraces all the numbers on both sides.

Now let us start transposing the numbers. The only way to remove 2 in the divisor on the LHS is to multiply both LHS and RHS numerators by 2. This allows you to cancel the 2 on the LHS.

You must remember that there is always an invisible 1 multiplier associated with a single number, symbol or term (and other terms) and when the number is totally removed, you are left with 1:

$$\frac{2 \times 4[60 - 40 + (2 \times 5)]}{2} = \frac{60 \times 2}{1}$$

Whilst you cannot transpose any of the numbers inside the brackets you can cancel out 2 on the LHS. You can also transpose 4 to the denominator on the RHS which is the result of dividing both sides by 4:

$$\frac{4\,[60 - 40 + (5 \times 2)]}{4} = \frac{60 \times 2}{4} = 30$$

Cancelling out 4 on the LHS leaves:

$$60 - 40 + 10 = 30$$

With no numbers as divisors, you can equate the problem to zero and it will demonstrate to you how the numbers change their sign when they are moved directly across the equal sign. Thus by adding −60 to both sides you obtain:

$$-60 + 60 - 40 + 10 = 30 - 60$$

leaving

$$-40 + 10 = 30 - 60$$

By adding +40 to both sides you obtain:

$$40 - 40 + 10 = 30 - 60 + 40$$

leaving

$$10 = 30 - 60 + 40$$

By adding −10 to both sides you obtain:

$$-10 + 10 = 30 - 60 + 40 - 10$$

leaving

$$0 = 30 - 60 + 40 - 10$$

Note: Having finished this example, it is worth remembering that when a number is multiplied by 0 the answer is 0. If it is divided by 0 the answer is infinity and when 0 is divided by the number the answer is 0.

Square root and cube root signs

The square root of a number is the number whose square equals the given number. For example, $\sqrt{25} = 5$ (and $5^2 = 25$). Also note that $\sqrt{1} = 1$ since $1^2 = 1$.

The cube root of a number is the number whose cube equals the given number. For example, $\sqrt[3]{125} = 5$ (and $5^3 = 125$).

When formulae involve root signs, try and simplify the task by removing the root sign first. For example:

If

$$\sqrt{abc} = d$$

then

$$abc = d^2$$

If

$$r = \sqrt{\frac{A}{\pi}}$$

then

$$r^2 = \frac{A}{\pi}$$

Now consider transposition of formulae using algebraic quantities. In the last example you saw that:

$$r^2 = \frac{A}{\pi}$$

Now make A the subject of the formula. Start by removing π to the numerator on the LHS leaving A the subject. Thus:

$$\pi r^2 = A$$

It is generally accepted that the subject of a formula is written on the LHS allowing the other quantities to be expressed on the RHS, i.e:

$$A = \pi r^2$$

You should try and get into the habit of writing down 'introductory' words for each step taken in evaluating the formula, e.g. if, since, then, and, therefore, also, hence (from here), thus (as a result of this). Consider the following example:

If

$$S = \frac{\sqrt{I^2 t}}{k}$$

To make t the subject of the formula, firstly, move k to the numerator on the LHS, then squre both sides and remove I to the denominator on the LHS:

$$Sk = \sqrt{I^2 t}$$

and

$$(Sk)^2 = I^2 t$$

therefore

$$\frac{(Sk)^2}{I^2} = t$$

or

$$t = \frac{S^2 k^2}{I^2}$$

Example 1.7

Make A, ρ, and l subjects of the formula:

$$R = \frac{\rho l}{A}$$

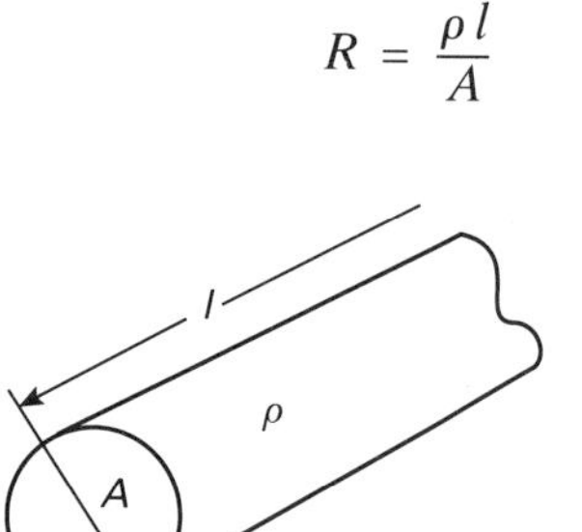

Resistance (R) depends on a conductor's properties

Figure 1.8 Resistance factors

Solution

$$A = \rho l/R$$

$$\rho = RA/l$$

$$l = RA/\rho$$

Example 1.8

Make R and V subjects of the formula:

$$I = \frac{V}{R}$$

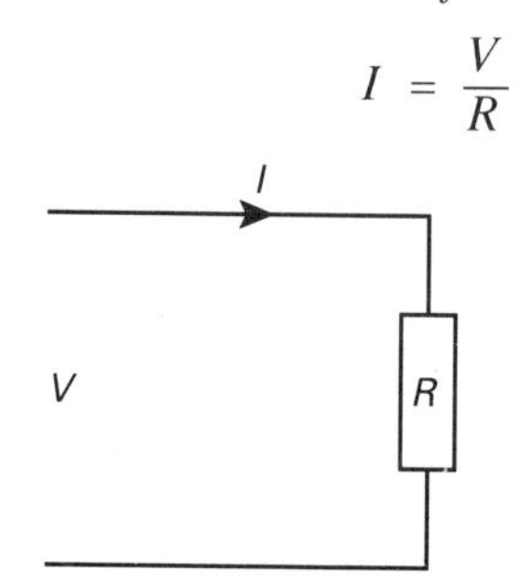

Current (I) is driven around circuit by source voltage but limited by resistance (R)

Figure 1.9 Limiting current flow

Solution

$$R = V/I \quad \text{and} \quad V = IR$$

Example 1.9

Make E, R and r subjects of the formula:

$$V = \frac{E}{R + r}$$

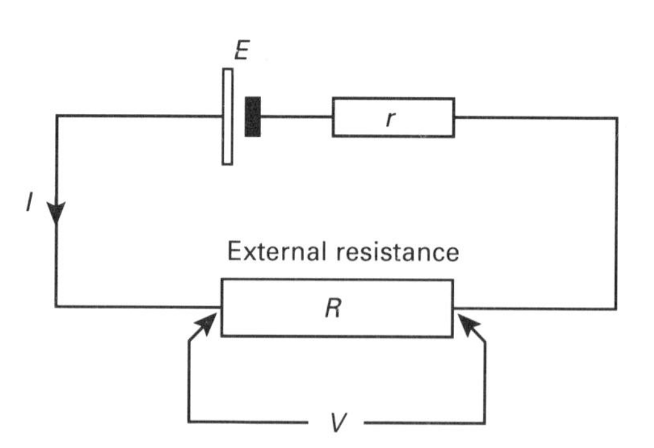

A battery has internal resistance (r)

Figure 1.10 Internal and external circuit resistance

Solution

$$E = V(R + r)$$

$$R = (E/V) - r$$

$$r = (E/V) - R$$

Example 1.10

Make R_1 subject of the formula:

$$Z_S - Z_E = (R_1 + R_2)L$$

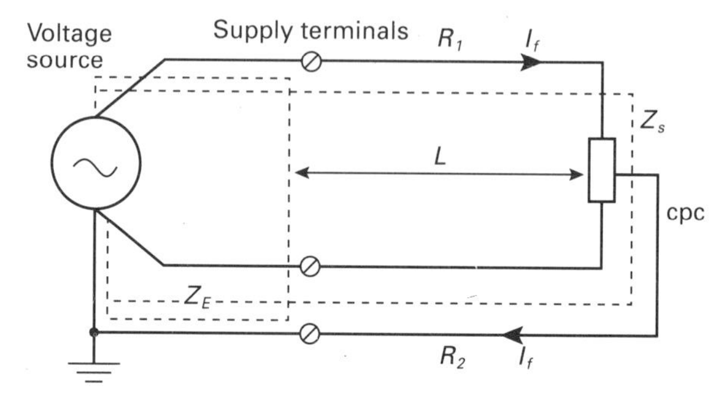

Fault current (I_f) flows through different elements of a circuit

Figure 1.11 System impedance

Solution

$$R_1 = [(Z_S - Z_E)/L] - R_2$$

Example 1.11

Make R the subject of the formula:

$$Z = \sqrt{R^2 + X^2}$$

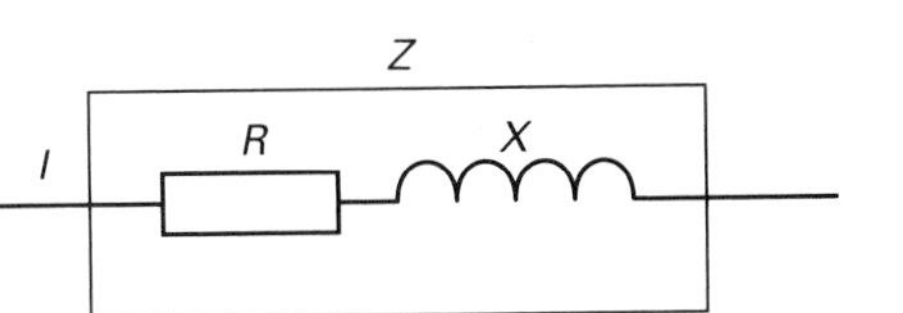

Figure 1.12 Impedance

Solution

$$R = \sqrt{Z^2 - X^2}$$

Example 1.12

If $X_L = 2\pi fL$ and $X_C = 1/2\pi fC$ make f the subject of the formula:

$$X_L = X_C$$

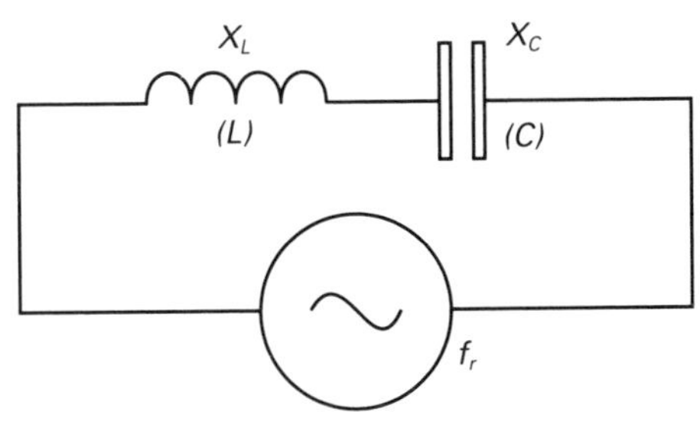

Figure 1.13 Circuit resonance

Solution

$$f = \frac{1}{2\pi\sqrt{LC}}$$

Example 1.13

Make N_R the subject of the formula:

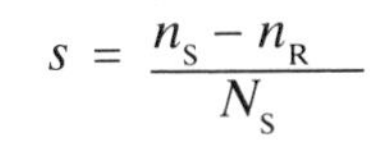

$$s = \frac{n_S - n_R}{N_S}$$

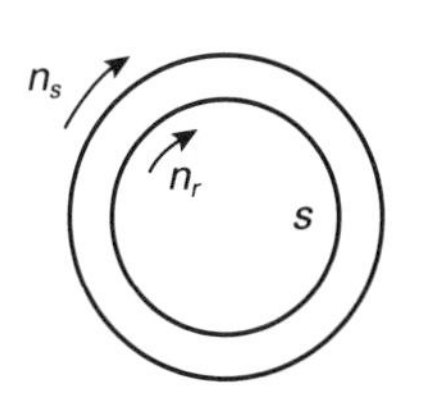

Slip (s) is the difference between synchronous speed (n_s) and rotor speed (n_r)

Figure 1.14 Induction motor

Solution

$$n_R = n_S(1 - s)$$

Example 1.14

Make V the subject of the formula:

$$s = \frac{V^2 t}{R}$$

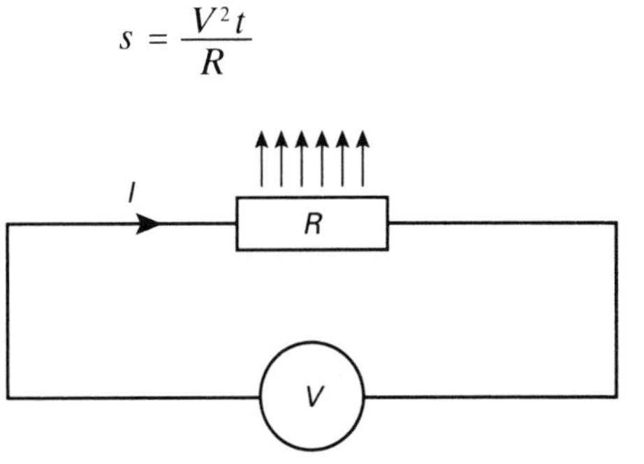

A resistor (R) produces heat energy (W)

Figure 1.15 Heat energy

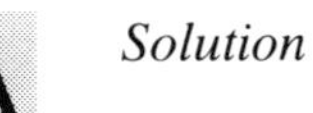

Solution

$$V = \sqrt{WR/t}$$

Example 1.15

Make R_2 the subject of the formula:

$$\frac{1}{R_E} = \frac{1}{R_1} + \frac{1}{R_2}$$

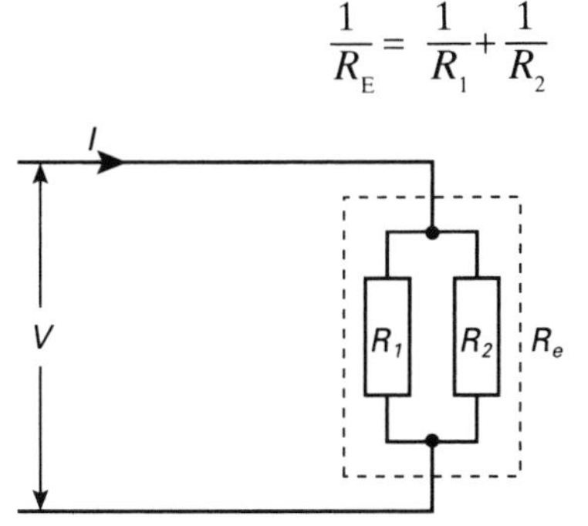

Equivalent resistance (R_e) in a parallel circuit has an ohmic value smaller than any circuit resistor

Figure 1.16 Parallel resistor connections

Solution

$$R_2 = R_1 R_E/(R_1 - R_E)$$

Example 1.16

Make n the subject of the formula:

$$R = \frac{nE}{R + nr}$$

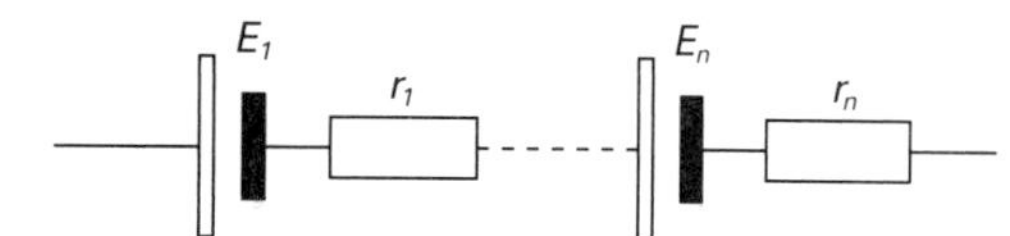

The more cells connected in series the higher the voltage and internal resistance

Figure 1.17 Connection of cells in series

A *Solution*

$n = IR\ (Ir - E)$

Scientific calculator

The course 236-8 Electrical Competence syllabus requires students to use a calculator to perform the four basic operations of arithmatic. These operations along with other techniques such as percentages, square roots, memory calculations and trigonometric functions will be dealt with in this section. In order for you to become a proficient user of your calculator, you must be conversant with its keyboard functions and you must acquire plenty of practice using it. Figure 1.18 shows the layout of a typical keyboard with second functions omitted for clarity. Listed below is a brief explanation of the main operating keys.

- **2ndF** key is used when you want to perform second operating functions with selected keys (those identified by smaller print and often a different colour).
- **ln** key and **log** key are used to calculate logarithms.
- **OFF** key is used to switch the power 'off' and normally has a facility to do this automatically after an interval of time if no other keys have been used.
- **ON/C** key is used to switch the power 'on' and is also the facility to cancel keying errors input.
- **sin**, **cos** and **tan** keys are used for trigonometric calculations.
- **F–E** key is used for changing the display mode.
- **CE** key is used to clear an entry.
- **Exp** key allows entry of exponential numbers in standard form.
- **x^y** key is used to perform power calculations, giving the value of x to the index y.
- **x^2** key is used to calculate the square of a number.
- **(** and **)** keys are used with bracket calculations.

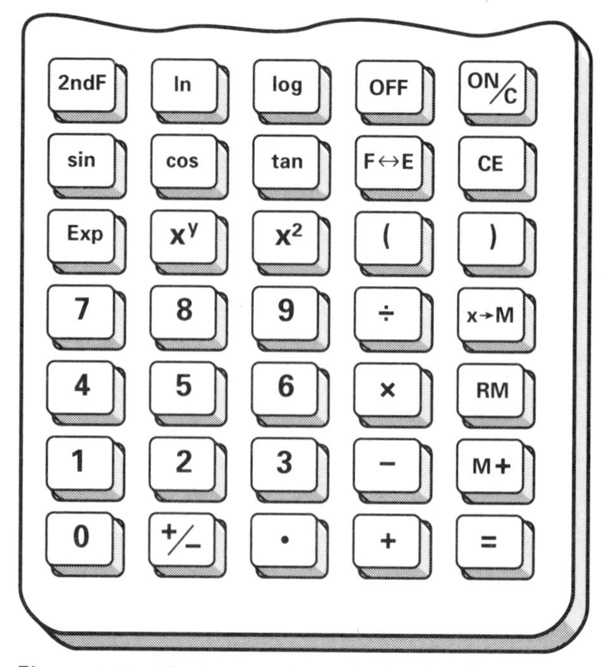

Figure 1.18 Keyboard of a typical scientific calculator

(Second functions have been omitted for clarity)

- **0 to 9** are number keys with the decimal point key shown as a dot.
- **+/–** key is used to change the sign of a displayed number.
- **equal sign (=)** key is used to obtain a calculation result.
- **x→M** key is used to store a number in the calculator.
- **RM** key is used to recall a stored number.
- **M+** key is used to add a further entry to an entry already stored in the memory.

Until you are absolutely conversant with your calculator, it is important that you make a rough check of any calculation. The golden rule to remember is that any error keyed is an error creating a wrong answer. The following worked examples are provided to help your understanding of the keyboard functions. The words 'press' and 'obtain' are action points with the latter word making the assumption that you are able to tap into your calculator whole and part numbers.

Example 1.17

Evaluate:

$$\frac{5.5 \times 6.2 \times 19.99}{5.084 \times 12.1 \times 7.8}$$

Solution

A rough check would express the numerator as 6 × 6 × 20 = 720 and the denominator as 5 × 12 × 8 = 480. Hence: 720 ÷ 480 = 1.5.

It should be noted that numbers containing decimal fractions of 0.5 and above have been rounded off to the next highest numbers. There are no definite rules for obtaining an approximate answer but rounding off numbers in order to cancel out on either side of the quotient line greatly simplifies matters.

Steps in using your calculator are:

1. press **ON/C** key
2. obtain **5.5**
3. press **×** key
4. obtain **6.2**
5. press **×** key (display shows **34.1**)
6. obtain **19.99**
7. press **=** key (display shows **681.659**)
8. press ÷ key
9. obtain **5.084**
10. press ÷ key (display shows **134.07927**)
11. obtain **12.1**
12. press ÷ key (display shows **11.080931**)
13. obtain **7.8**
14. press **=** key (display shows **1.4217064**)

Answer = 1.42 (to three significant figures).

Note: Significant figures are counted from left to right, starting with the first non–zero figure. In general, answers are accepted to two decimal points.

As an alternative method, you could multiply the denominator completely out and place the answer 479.82792 in the memory using the memory enter **x→M** key. If you now multiply the numerator completely out to obtain 681.659, then press the ÷ key, recall memory **RM** key and the equal sign **=** key, the same answer should appear.

Example 1.18

Evaluate:

$$\frac{0.03 \times 18.5 \times 10^2}{88.8 \times 0.09 \times 10}$$

Solution

Rough check:

$$\frac{0.03 \times 20 \times 100}{90 \times 0.09 \times 10} = \frac{60}{81} = 0.74$$

Steps using your calculator are:

1. press **ON/C** key
2. obtain **0.03**
3. press **×** key
4. obtain **18.5**
5. press **×** key
6. obtain **100**
7. press **=** key (display shows **55.5**)
8. press ÷ key
9. obtain **88.8**
10. press ÷ key (display shows **0.625**)
11. obtain **0.09**
12. press ÷ key (display shows **6.944**)
13. obtain **10**
14. press **=** key (display shows **0.694**)

Answer = 0.69

Example 1.19

Find the current taken by a 415 V/20 kW three-phase induction motor if it has a power factor of 0.7 lagging and an efficiency of 89%. The current is found from the formula:

$$I_L = (P \times 100) \div (\sqrt{3} \times V_L \times \text{p.f.} \times \%\ \text{effy})$$

$$\frac{20\,000 \times 100}{1.732 \times 415 \times 0.7 \times 89}$$

Solution

Rough check:

$$\frac{2\,000\,000}{2 \times 400 \times 0.5 \times 100} = \frac{200}{4}$$

Steps using your calculator are:

1. obtain **2 000 000**
2. press ÷ key
3. obtain **1.732**
4. press ÷ key (display shows **1154734.4**)
5. obtain **415**
6. press ÷ key (display shows **2782.4926**)
7. obtain **0.7**

8. press ÷ key (display shows **3974.9894**)
9. obtain **89**
10. press **=** key
    (display shows answer **44.662802**)

Answer = 44.66 A

Note: The square root of 3 (i.e. √3) = 1.732. For this function, press number 3 key and then press the **2ndF** key and the $\mathbf{x^2}$ key.

If there are more numbers in the denominator than in the numerator, it would be easier and quicker to multiply the numbers in the denominator first, place the answer obtained in the memory by pressing the **x→M** key, then tap in the numerator numbers. You should then press the divide key followed by the recall memory **RM** key and obtain your answer by pressing the equal sign key.

Example 1.20

Find the disconnection time (t) for a fuse that protects a 50 mm² twin pvc-insulated cable (S) having copper conductors, if its insulation material factor (k) is taken as 115 and its fault level (I_F) estimated to be 4000 A. The formula used is:

$$t = (k^2 \times S^2) \div I_F^2$$

Solution

Hence:

$$t = (115^2 \times 50^2) \div 4000^2$$

Your rough answer check is:

$$t = \frac{100 \times 100 \times 50 \times 50}{4000 \times 4000} = \frac{25}{16}$$

Steps using your calculator are:

1. press **ON/C** key
2. obtain **115**
3. press $\mathbf{x^2}$ key (display shows **13,225**)
4. press **×** key
5. obtain **50**
6. press $\mathbf{x^2}$ key
7. press **=** key
   (display shows **3,306,2500**)
8. press ÷ key
9. obtain **4,000**
10. press $\mathbf{x^2}$ key
11. press **=** key
    (display shows **2.0664063**)

Answer = 2.07 s

Example 1.21

Find the cross sectional area (A) of an insulated cable if it has an overall diameter (d) of 6.2 mm. Here, $A = \pi d^2/4$

Solution

$$A = (3.142 \times 6.2^2) \div 4$$

A rough check gives:

$$A = (3 \times 40) \div 4 = 30$$

Using your calculator:

1. press **ON/C** key
2. press **2ndF** key
3. press **EXP** key
   (display shows π = **3.1415927**)
4. press **×** key
5. obtain **6.2**
6. press $\mathbf{x^2}$ key
7. press **=** key (display shows **120.76282**)
8. press ÷ key
9. obtain **4**
10. press **=** key (display shows **30.1907**)

Answer = 30.19 mm²

Note: The divisor **4** could be inverted by pressing the **2ndF** key and then by pressing the **(** key. This changes 4 into 0.25 (its reciprocal). The calculation can now be completed in the numerator.

Example 1.22

Determine the resistance (R) of a 1.5 mm² twin and earth pvc-sheathed copper cable if its length (l) is 25 m (50 m overall) and its resistivity (ρ) is taken to be 0.0172 μΩm. The formula for finding resistance is

$$R = (\rho \times l) \div A$$

Solution

$$R = (0.0172 \times 50 \times 10^{-6}) \div (1.5 \times 10^{-6})$$

Rough check

$$R = (0.02 \times 50 \times 10^{-6}) \div (2 \times 10^{-6})$$

$$= 0.5$$

Using your calculator:

1. press **ON/C** key
2. obtain **0.0172**
3. press **×** key
4. obtain **50**
5. press **=** key (display shows **0.86**)
6. press **÷** key
7. obtain **1.5**
8. press **=** key (display shows **0.57333**)

Answer = 0.57 Ω

Note: The above steps have considered the cancelling out of 10^{-6} since it appears in both numerator and denominator.

Example 1.23

The formula for finding a cable's tabulated single-circuit current carrying capacity (I_t) is given by:

$$I_t = \sqrt{I_n^2 + 0.48 I_b^2 \times (1 - C_g^2)/C_g^2}$$

If I_b = 4.16 A, I_n = 6 A and C_g = 0.65, determine the value of I_t

Solution

Here, it is difficult to obtain a rough check but those involved in this type of cable selection should be able to see from the data that the required value of I_t will be slightly greater than the I_n value.

$$I_t = \sqrt{6^2 + (0.48 \times 4.16^2) \times (1 - 0.652)/0.65^2}$$

Using your calculator proceed as follows:

1. press **ON/C** key
2. obtain **0.65**
3. press **x²** key
4. press **+/–** key
5. press **+** key
6. press **1** key
7. press **=** key (display shows **0.5775**)
8. press **÷** key
9. obtain **0.65**
10. press **x²** key
11. press **=** key (display shows **1.3668639**)
12. press **×** key
13. obtain **4.16**
14. press **x²** key (display shows **17.3056**)
15. press **×** key
16. obtain **0.48**
17. press **=** key (display shows **11.3541**)
18. press **+** key
19. obtain **36**
20. press **=** key (display shows **47.3541**)
21. press **2ndF** key
22. press **x²** key (display shows **6.88143**)

Answer = 6.88 A

Example 1.24

Solve the expression:

$$5.55 \times 10^4 \div 9.81 \times 10^6$$

Solution

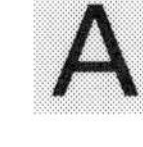

The steps on your calculator should be as follows:

1. press **ON/C** key
2. obtain **5.55**
3. press **EXP** key
4. obtain **4**
5. press **÷** key (display reads **55.500**)
6. obtain **9.81**
7. press **EXP** key
8. obtain **6**
9. press **=** key (display reads **0.0056574**)

Answer 0.006 = 6×10^{-3}

Example 1.25

(i) Solve

$$\frac{(0.727 + 1.15) \times (230 + 70) \times 27}{10^3 \times (230 + 20)}$$

(ii) Solve

$$\frac{1.6 \times 10^3 \times 8.64 \times 10^{-4}}{(6.76 \times 10^2) + (3.24 \times 10^{-2})}$$

Solution

Answer (i) 0.061 = 6.1×10^{-2}

Answer (ii) 0.002 = 2×10^{-3}

Note: When dealing with minus powers, remember to press the **+/–** key directly after the **EXP** key.

Example 1.26

Figure 1.19 shows a right-angled triangle representing impedance (Z), resistance (R)

and capactive reactance (X_c). Determine the unknown side using the formula.

$$X_c = \sqrt{Z^2 - R^2}$$

Solution

Answer X_c = 105.1 Ω

Example 1.27

Determine the rotor speed n_r of a 6-pole cage induction motor operating from a 50 Hz supply with a 7% slip.

The formulae to use are $n_r = n_s(1 - s)$ and $n_s = f/p$ where n_s is the synchronous speed of the stator, f is the frequency and p is the number of pole pairs (3).

Solution

Answer n_r = 15.5 rev/s

Other calculator applications

Percentages

Example 1.28

What is 4% of 240 V?

Solution

Press/obtain the following keys:

1. **ON/C**
2. **240**
3. **×**
4. **4**
5. **2ndF**
6. **%**
7. **=**

Answer = 9.6 V

Reciprocals

Example 1.29

A 20 Ω resistor and 60 Ω resistor are connected in parallel. What is their equivalent resistance?

Solution

Press/obtain the following keys:

1. **ON/C**
2. **20**
3. **2ndF**
4. **1/x**
5. **+**
6. **60**
7. **2ndF**
8. **1/x**
9. **=**
10. **2ndF**
11. **1/x**

Answer = 15 Ω

Trigonometry

Example 1.30

What are the sine, cosine and tangent of the following angles: 0^0, 30^0, 45^0, 60^0 and 90^0?

Solution

Press the following keys:

1. **0; sin;** (Ans=0)
2. **0; cos;** (Ans=1)
3. **0; tan;** (Ans=0)
4. **30; sin;** (Ans=0.5)
5. **30; cos;** (Ans=0.866)
6. **30; tan;** (Ans=0.577)

Repeat this procedure for the other angles.

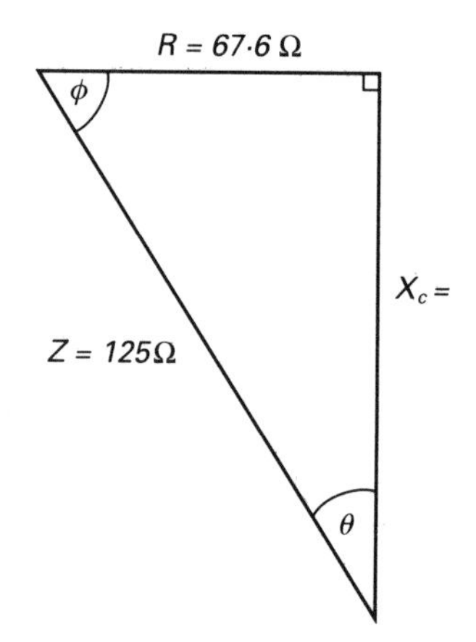

Figure 1.19 Impedance triangle

Example 1.31

In Figure 1.19 what are the angles created between the ratios *R/Z* and X_c/Z?

Solution

For angle ϕ, press/obtain the following keys:

1. **ON/C**
2. **67.6**
3. ÷
4. **125**
5. **=**
5. **2ndF**
6. **cos**

Answer $\phi = 57.262°$

For angle θ, press/obtain the following keys:

1. **105.1**
2. ÷
3. **125**
4. **=**
5. **2ndF**
6. **cos**

Answer $\theta = 32.775°$

Example 1.32

The formula required in Figure 1.20 is

$$X = \sqrt{S^2 - P^2}$$

If $X = 30\ \Omega$ and $S = 120\ \Omega$, find the value of *P* and the angle between the sides *X/S*. The ratio of the sides *P/S* is called the power factor (cos ϕ). Determine this condition using the values given.

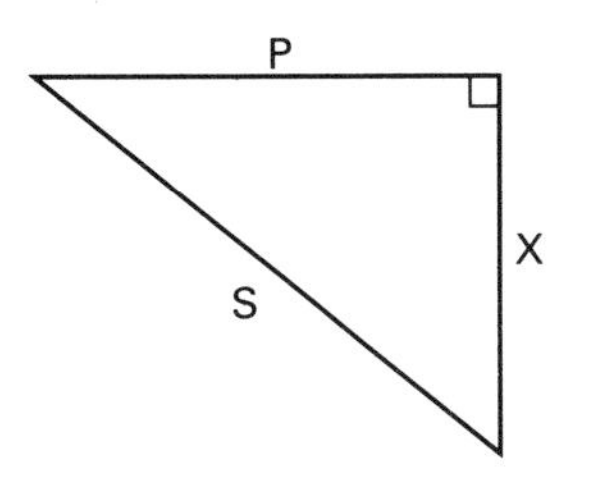

Figure 1.20 Power triangle

Solution

For the value *P*, press/obtain the following keys:

1. **ON/C**
2. **120**
3. **x^2**
4. **–**
5. **30**
6. **x^2**
7. **=**
8. **2ndF**
9. **x^2**

Answer $P = 116.2\ \Omega$

For the angle between *X/S*, press/obtain the following keys:

1. **ON/C**
2. **30**
3. ÷
4. **120**
5. **=**
6. **2ndF**

Answer $\theta = 75.52°$

For the ratio between P/S, press/obtain the following keys:

1. **ON/C**
2. **116.2**
3. ÷
4. **120**
5. **=**

Answer cos $\phi = 0.968$

Trigonometrical ratios

Figure 1.21 shows a right-angled triangle. The line facing the right angle is called the **hypotenuse** and the other two sides called **adjacent** and **opposite**. The two acute angles are given Greek letters θ (theta) and ϕ (phi) and you will see that the adjacent side for angle ϕ is the opposite side for angle θ. Similarly, the adjacent side for angle θ is the opposite side for angle ϕ. The trigonometrical ratios of the three sides are called: **sine**, **cosine** and **tangent** (abbreviated sin, cos and tan) and are expressed as follows:

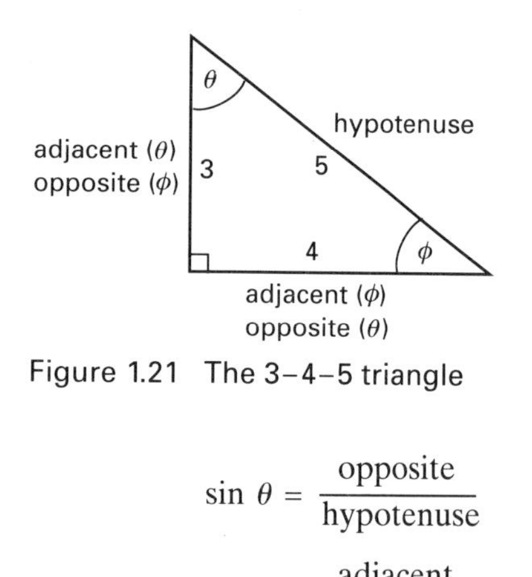

Figure 1.21 The 3–4–5 triangle

$$\sin\theta = \frac{\text{opposite}}{\text{hypotenuse}}$$

$$\cos\theta = \frac{\text{adjacent}}{\text{hypotenuse}}$$

$$\tan\theta = \frac{\text{opposite}}{\text{adjacent}}$$

$$\sin\phi = \frac{\text{opposite}}{\text{hypotenuse}}$$

$$\cos\phi = \frac{\text{adjacent}}{\text{hypotenuse}}$$

$$\tan\phi = \frac{\text{opposite}}{\text{adjacent}}$$

Since there are 180° in any triangle, ϕ and θ cannot exceed 90°. Their range of values may be found from trigonometrical tables or from a scientific calculator.

In Science 1 book, under the sub-heading of mensuration, the 3–4–5 right-angled triangle aptly illustrated the **theorem of Pythagoras** by showing that the square on the hypotenuse was equal to the sum of the squares on the other two sides (e.g. $3^2 + 4^2 = 5^2$). This triangle will be used to find the sine, cosine and tangent of the above trigonometrical ratios:

$$\sin\theta = \frac{3}{5} = 0.6$$

Therefore

$$\theta = 36.870$$

Also

$$\cos\theta = \frac{4}{5} = 0.8$$

$$\tan\theta = \frac{3}{4} = 0.75$$

$$\sin\phi = \frac{4}{5} = 0.8$$

And

$$\phi = 53.13^\circ$$

$$\cos\phi = \frac{3}{5} = 0.75$$

$$\tan\phi = \frac{4}{3} = 1.333$$

It will be seen that $\theta + \phi = 36.87^\circ + 53.13^\circ = 90^\circ$. It should be noted that both angles express the degrees as a whole number and part number. To convert the decimal part into minutes you simply multiply by 60, i.e. $0.87 \times 60 = 52.2$ mins and $0.13 \times 60 = 7.8$ mins.

Useful hints

For tangents of angles less than 90°

- when the angle is 0°, tan 0° = 0
- when the angle increases the tangent increases
- when the angle is 45°, tan 45° = 1
- when the angle approaches 90°, tan 90° approaches infinity.

For sines of angles less than 90°

- when the angle is 0°, sin 0° = 0
- when the angle increases the sine increases
- when the angle is 90°, sin 90° = 1

For cosines of angles less than 90°

- when the angle is 0°, cos 0° = 1
- when the angle increases the cosine decreases
- when the angle is 90°, cos 90° = 0

Ratios of common angles

Figure 1.22 shows an equilateral triangle (all sides and angles equal). The length of CD is found by: sin 60° = CD/AC.

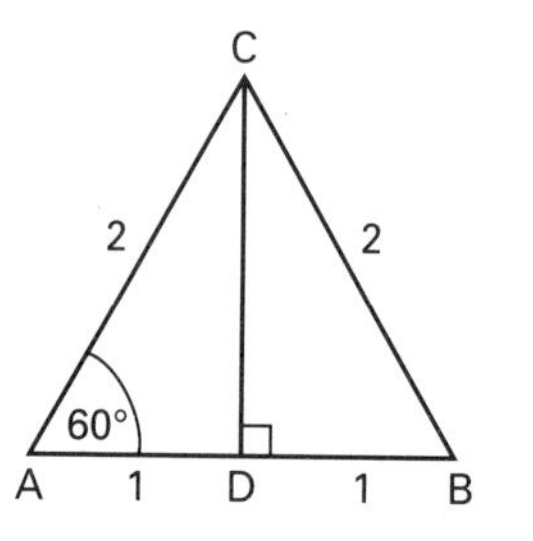

Figure 1.22 Equilateral triangle

Therefore

$$CD = \sin 60° \times AC$$
$$= 0.866 \times 2$$
$$= \sqrt{3}$$

hence

$$\sin 60° = \sqrt{3}/2$$
$$\tan 60° = \sqrt{3}/1$$
$$\cos 60° = 1/2$$

also

$$\sin 30° = 1/2$$
$$\tan 30° = 1/\sqrt{3}$$

and $$\cos 30° = \sqrt{3}/2$$

Figure 1.23 shows an isosceles triangle (two sides and two angles equal) where $\cos 45° = BC/AC$.

Therefore

$$AC = BC/\cos 45°$$
$$= 1/0.707$$
$$= 1.414$$
$$= \sqrt{2}$$

hence:

$$\cos 45° = 1/\sqrt{2}$$
$$\sin 45° = 1/\sqrt{2}$$

and $$\tan 45° = 1$$

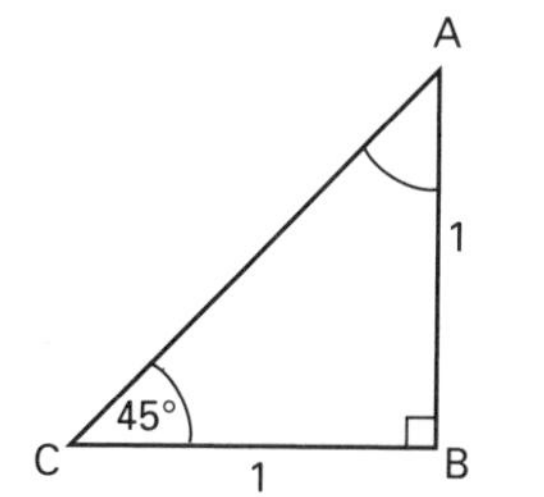

Figure 1.23 Isosceles triangle

Complementary angles

These are angles whose sum is 90°. The sine of an angle is equal to the cosine of its compliment, and the cosine of an angle is equal to the sine of its compliment. You will see from Figure 1.21 that:

$$\sin \theta = \frac{\text{opposite}}{\text{hypotenuse}}$$

and also

$$\cos \theta = \frac{\text{adjacent}}{\text{hypotenuse}}$$

Hence $\sin \theta = \cos \phi$

and $\sin \theta = \cos (90° - \theta)$

also $\cos \theta = \sin (90° - \theta)$

Relations between ratios

This can be explained with reference to Figure 1.24

Here

$$\sin \theta = \text{opposite (AC)/hypotenuse (AB)}$$

and

$$\cos \theta = \text{adjacent (BC)/hypotenuse (AB)}$$

therefore

$$\sin \theta/\cos \theta = (AC/AB)/(BC/AB)$$
$$= (AC/AB) \times (AB/BC)$$
$$= AC/BC$$
$$= \tan \theta$$

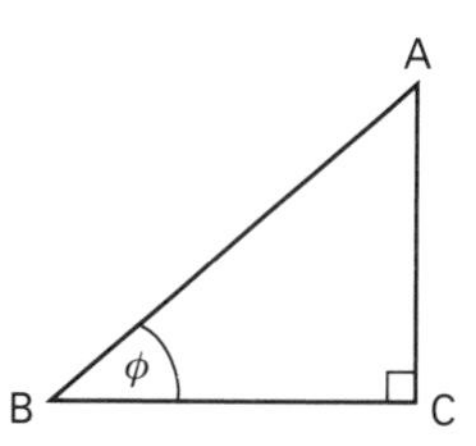

Figure 1.24 The general case for a right-angled triangle

Q

Example 1.33

With reference to Figure 1.26 find the side marked X and the two acute angles θ and ϕ.

A

Solution

Since

$$\cos \phi = \text{adjacent/hypotenuse}$$
$$= 10/20$$
$$= 0.5$$

therefore

$$\phi = 60°$$

also

$$\theta = 90° - \phi$$
$$= 90° - 60°$$
$$= 30°$$

Since

$$\sin \phi = \text{opposite/hypotenuse}$$
$$= X/20$$

then

$$X = \sin \phi \times 20$$
$$= \sin 60° \times 20$$
$$= 0.866 \times 20$$
$$= 17.32$$

Note: This example should be carried out using a scientific calculator in they way previously described. For example, to obtain angle ϕ:

1. press **ON/C** key
2. divide **10** by **20**
3. press **=** key (display shows **0.5**)
4. press **2ndF** key
5. press **cos** key (display shows **60°**)

To obtain side X

1. press **sin** key (display shows **0.866**)
2. multiply **0.866** by **20**
3. press **=** key (display shows **17.32**)

You can solve for X simply by applying Pythagorus' theorem:

$$X = \sqrt{20^2 - 10^2}$$
$$= 17.32$$

You could also find X by using the ratio tan θ = opposite/adjacent.

Solution of triangles

As seen above, every triangle consists of three sides and three angles but for triangles which are not right-angled the sine rule and cosine rule can be used.

Sine rule

This rule is used when you are given one side and any two angles or two sides and an angle opposite to one of the sides. Figure 1.25 shows an acute-angled triangle with its angles labelled A, B, and C and sides labelled a, b, and c. If a line AD (side h) is drawn perpendicular to BC, then in angle ACD:

$$\sin C = h/b$$

therefore

$$h = \sin C \times b \qquad [1]$$

and in angle ABD:

$$\sin B = h/c$$

therefore

$$h = \sin B \times c \qquad [2]$$

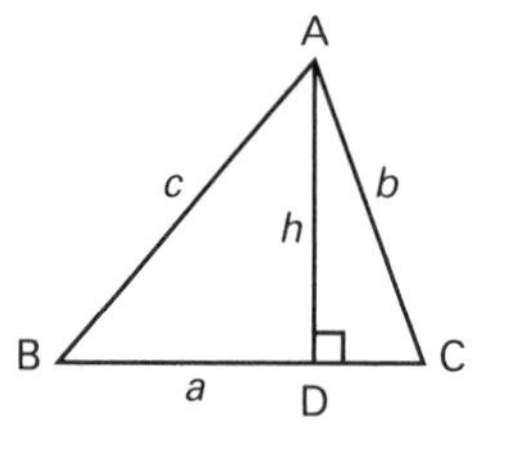

Figure 1.25 The sine rule

Since [1] and [2] are equal:

$$\sin C \times b = \sin B \times c$$

Therefore

$$\frac{b}{\sin B} = \frac{c}{\sin C}$$

In a similar way it is possible to find the third ratio:

$$\frac{a}{\sin A}$$

Any pair of ratios may be used to solve a triangle for the conditons mentioned.

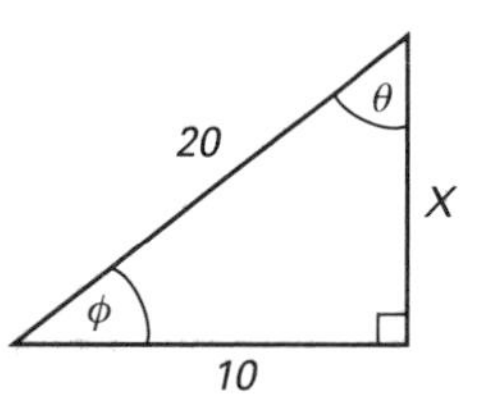

Figure 1.26 Right-angled triangle

Cosine rule

This rule is used when you are given two sides of a triangle and the angle between them or if you are given three sides of a triangle. Its proof will not be given but it is stated as:

$$a^2 = b^2 + c^2 - 2bc \cos A$$

$$b^2 = a^2 + c^2 - 2ac \cos B$$

and

$$c^2 = a^2 + b^2 - 2\,ab \cos C$$

Q

Example 1.34

With reference to Figure 1.27, find the lengths of the unknown sides using a scientific calculator.

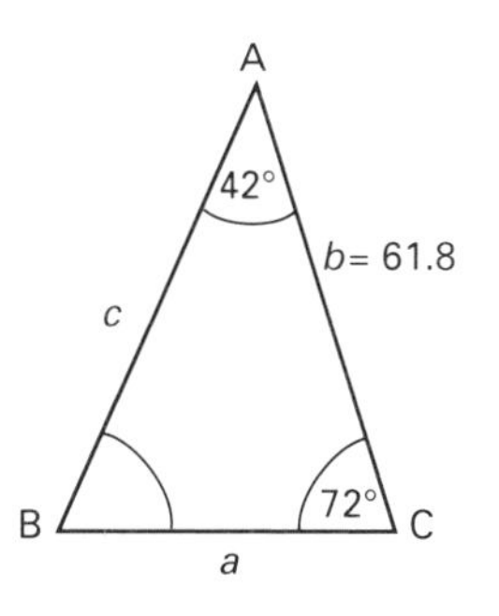

Figure 1.27 Example triangle

A

Solution

Express the sine rule formula to find side a, i.e.

$$a = \frac{b \times \sin A}{\sin B}$$

1. press **ON/C** key
2. obtain **66°** (unknown angle for B)
3. press **sin** key (display shows **0.9135**)
4. press **M+** key
5. obtain **42°** (angle A)
6. press **sin** key (display shows **0.6691**)
7. press × key
8. obtain **61.8**
9. press = key (display shows **41.3522**)
10. press ÷ key
11. press **RM** key (display shows **0.9135**)
12. press = key
    (display shows a to be **45.2656**)

To find side c, rearrange the sin rule to become:

$$c = \frac{b \times \sin C}{\sin B}$$

1. obtain **61.8**
2. press × key
3. obtain **72°** (angle C)
4. press **sin** key (display shows **0.951**)
5. press = key (display shows **58.775**)
6. press ÷ key
7. obtain **66**
8. press **sin** key
9. press = key
    (display shows c to be **64.337**)

Q

Example 1.35

Using your calculator, find the angles of the triangle shown in Figure 1.28.

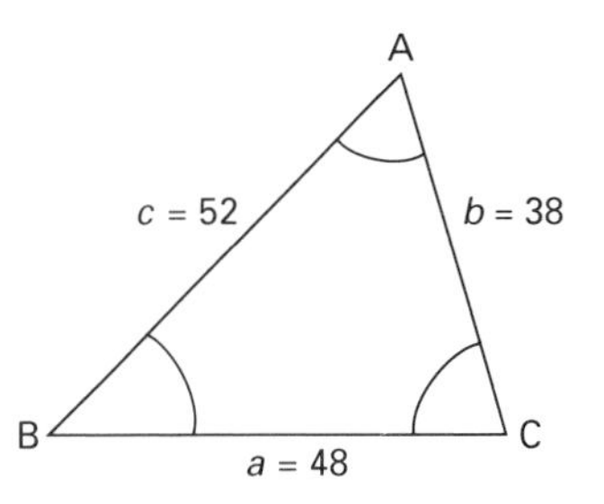

Figure 1.28 Example triangle

A

Solution

The solution to this formula is found by using the cosine rule. Use your calculator to confirm the following working:

$$\cos A = \frac{b^2 \times c^2 - a^2}{2bc}$$

Hence

$$\cos A = \frac{1444 + 2704 - 2304}{3952}$$

$$= 0.4666$$

therefore

$$A = 62.18°$$

You can now apply the sine rule to find angle B:

$$\sin B = \frac{\sin A \times b}{2bc}$$

Hence

$$\sin B = \frac{\sin 62.18° \times 38}{48}$$

$$= 0.7$$

therefore

$$B = 44.43°$$

The angle C is found by:

$180° - 62.18° - 44.43° = 73.39°$

EXERCISE 1.2

Simple equations

1. Solve for y when $7y + 10 = 4y + 19$
2. Solve for d when $12d - 5(d - 1) = 2d + 6$
3. Solve for n when $6 - 4(3 - n) = 3(n - 7)$
4. Solve for a when $2(a - 13) = 10 - 3(a + 2)$
5. Solve for x when $\dfrac{6}{x - 30} = \dfrac{4}{x + 15}$

Calculator problems

6. $9.18 \times 10^{-4} - 7.38 \times 10^{-6}$
7. $5.89 \times 10^{-3} + 2.68 \times 10^{-2}$
8. $2.65 \times 10^{3} + 8.17 \times 10^{-3}$
9. $6.89 \times 10^{5} + 4.73 \times 10^{3}$
10. $1.98 \times 10^{5} \times 4.65 \times 10^{3}$

Simultaneous equations

11. Solve $3x + 2y = 4$ [1]
    $x + 2y = 0$ [2]

12. Solve $3x + 2y = 9$ [1]
    $2x + 3y = 16$ [2]

13. Solve $8m - 3n = 39$ [1]
    $7m + 5n = -4$ [2]

14. Solve $3p - 2q = 7$ [1]
    $7p - 3q = 18$ [2]

15. Solve $6I_1 + 10I_2 = 2$ [1]
    $9I_1 + 6I_2 = 12$ [2]

Trigonometry

16. With reference to Figure 1.29 determine sides AC and CB.

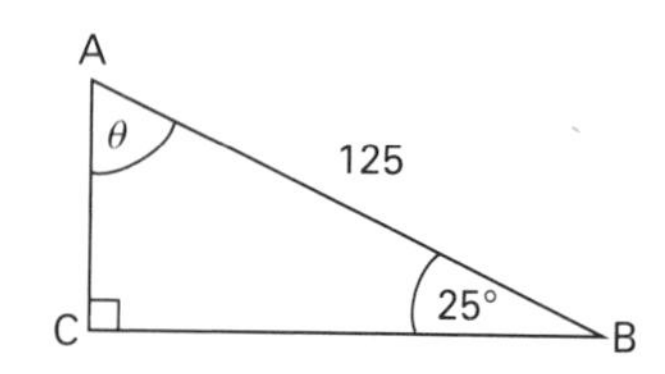

Figure 1.29

17. The power factor of an inductive circuit is 0.75 lagging. If the impedance is 30 Ω, what is the resistance and reactance of the circuit?

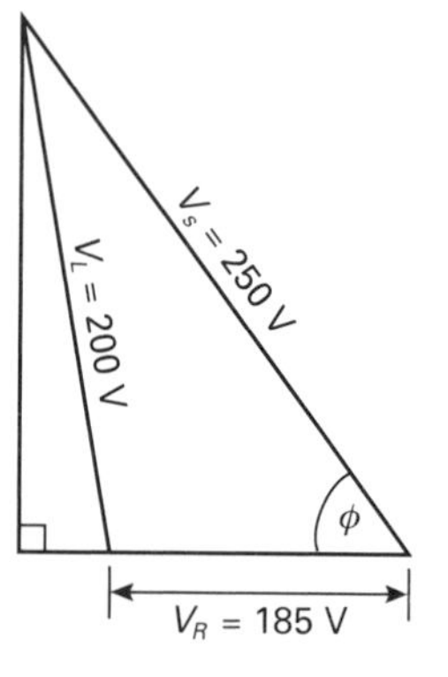

Figure 1.30

18. The reactive voltamperes of a circuit is 150 kVAr. If the apparent power is 210 kVA, what is the phase angle and true power of the circuit?

19. Figure 1.30 shows voltmeter readings across a resistor (V_R) an inductor (V_L) (containing resistance) and the a.c. supply (V_S). Determine the phase angle (ϕ).

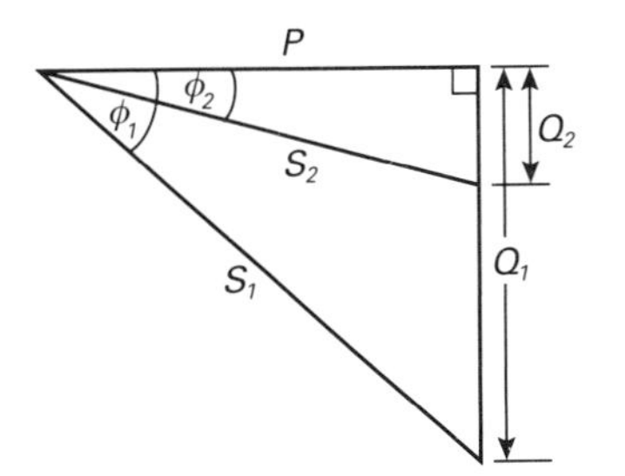

Figure 1.31

20. Figure 1.31 shows a power triangle.
If $P = 25$ kW and $\phi_1 = 53°$, determine S_1 and Q_1. If the phase angle decreases to 24° (ϕ_2), what is the value of S_2 and difference between Q_1 and Q_2?

Alternating current circuits

Objectives

After reading this chapter you should be able to:

- *describe the production and nature of a sinusoidal waveform;*
- *describe phase displacement between voltage and current waveforms for different circuit components;*
- *construct phasor diagrams of a.c. voltage and current quantities for different circuit components;*
- *state the meaning of power factor and show how it can be obtained from the construction of right-angled triangles;*
- *describe the production and nature of a three-phase system;*
- *state the need for a neutral conductor in a three-phase system to create phase balancing;*
- *state the difference between star and delta in relation to line and phase voltage and current;*
- *perform calculations to find power factor, true power, apparent power and reactive power in a.c. circuits.*

The sine wave

The generation of alternating current electricity is by **electromagnetic induction** created by an a.c. generator's rotating magnetic field cutting through stationary stator conductors. The voltage produced is cyclic in nature and Figure 2.1 shows how one cycle is produced when the generator's rotor makes one complete revolution. The shape of the graph is **sinusoidal** (i.e. of a sine wave) and you will see that it reaches two maximum values, one in a positive direction and the other in a negative direction. It also passes through zero twice in every cycle which is the point of the graph where the fastest rate of change occurs from one peak to the other.

The time taken for one cycle to occur is called the **periodic time** (T) and the number of cycles per second is called the **frequency** (f). The unit of frequency is called the **hertz** (Hz). Periodic time is expressed as:

$$T = 1/f \qquad [2.1]$$

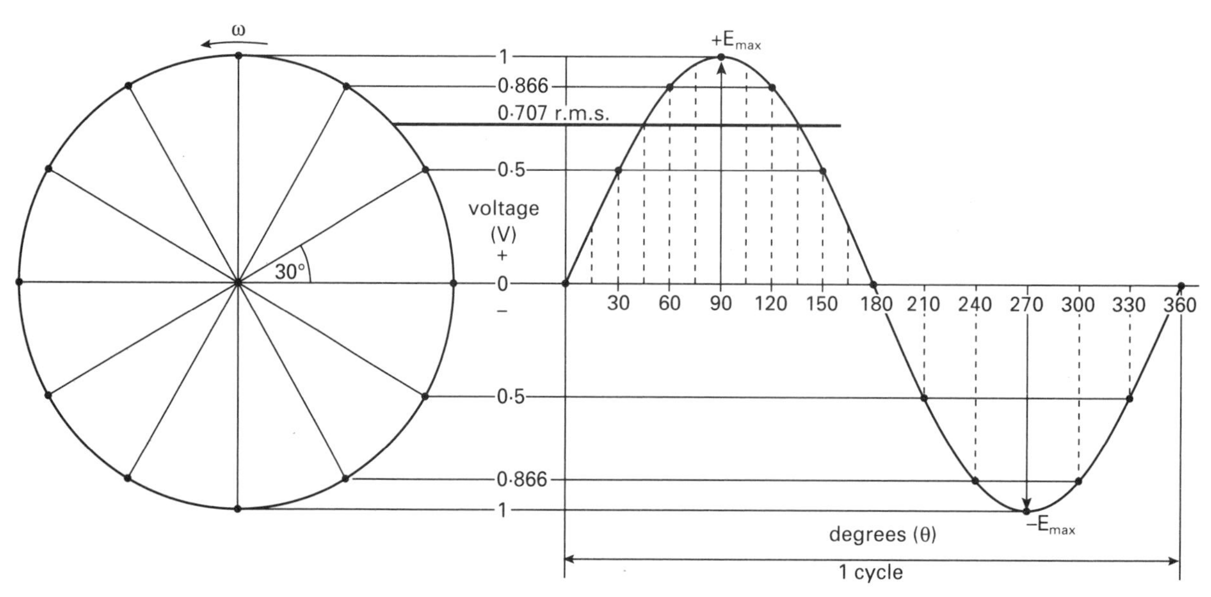

Figure 2.1 Generation of a sine wave

The public electricity supply in Britain is generated at a frequency of 50 Hz, which is 50 cycles per second and this means that the a.c. generator's rotor (n_r) with its travelling magentic field has to be driven at a constant 50 revs/s or 3 000 revs/min. The formula for finding the generator's rotor speed is:

$$n_r = f/p \qquad [2.2]$$

where p is the generator's magnetic field pole pairs.

It is important to note that when dealing with ordinary induction motors (asynchronous types) the same formula is used for finding synchronous speed (n_s) of the travelling magnetic field. The rotor speed of such motors is not the same since it has to consider slip.

In Figure 2.1 the line on the graph marked 0.707 is the effective value or **root mean square value** (r.m.s.). All a.c. supply voltages and currents are measured using r.m.s. values. The peak is called the **maximum value** and if the root mean square value is known the maximum value can easily be found by dividing it by 0.707. For example, if the supply voltage were 240 V the maximum value would be 240/0.707 = 339 V.

Circuit components

To understand a.c. theory, you need to know the current and voltage relationship of three circuit properties, namely **resistance** (R) the property of a resistor, **inductance** (L) the property of an inductor and **capacitance** (C) the property of a capacitor. Electrical equipment may possess one or more of these properties and it is usually the inductive component which is found to be the main cause of **phase difference** or shift between current and voltage quantities (see Figure 2.2). If there is no inductance in the circuit, the phase difference is usually zero, with the current rising and falling in phase with the voltage. If the phase difference is too great due to inductance in the circuit, it may affect the supply system and lead to higher installation and running costs for the consumer.

In Figure 2.2 the phase difference is represented by a **phase angle** (ϕ) and it is the cosine of this phase angle (cos ϕ) which is called **power factor** (*p.f.*) In a.c. circuits, power (P) is not just the product of voltage (V) and current (I) (as in d.c. circuits), it has to include power factor. The formula is expressed as:

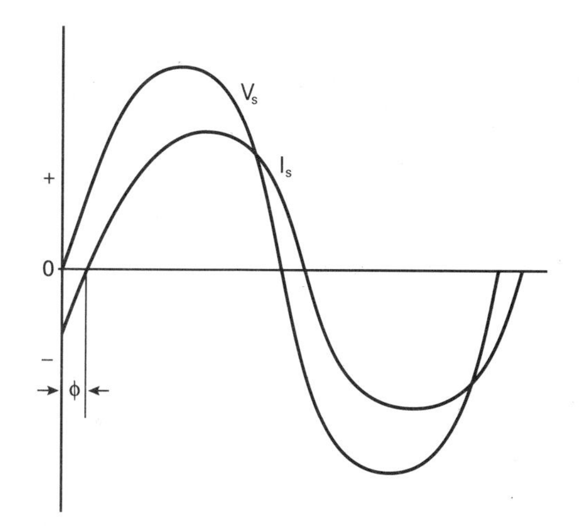

Figure 2.2 Phase displacement between supply voltage and supply current

$$P = P \times I \times p.f. \qquad [2.3]$$

By transposition

$$p.f. = P/VI \qquad [2.4]$$

Power factor has no unit like ampere for current or watt for power. If there is no phase angle, cos 0° = 1 (called **unity power factor**) and if it is 90° then cos 90° = 0. These are the two extreme conditions of power factor and if you insert either of these values in formula 2.3, you will see that power can only be consumed in circuits which create phase angles below 90°.

Resistance

When a **resistor** is connected to an a.c. supply (see Figure 2.3), it presents the only common opposition to current flow (i.e no inductance or capacitance). You will see from the graph that both current and voltage waveforms pass through the same instantaneous points together. This means that there is no phase angle between them and the quantities are said to be **in-phase** with each other. As already explained, if there is no phase angle then unity power factor exists. The **phasor diagram**, representing a rotating vector, indicates the magnitude and direction of the voltage and current quantities. Its rotation is in an anticlockwise direction and is normally drawn horizontally, making one of the quantities a reference. The two quantities are meant to be superimposed on each other but are drawn side by side for clarity. Either quantity can be drawn longer than the other, depending on

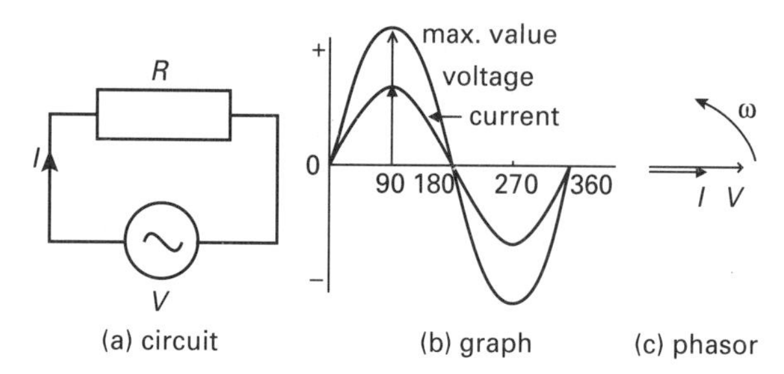

Figure 2.3 Circuit, graph and phasor diagram for 'purely' resistive a.c. circuit

scale, and to distinguish between them the voltage is given an open arrow and the current a closed arrow.

If the instantaneous values of voltage and current are multiplied together the resulting power curve would look like Figure 2.4(a). This clearly shows that a resistive component consumes power since there is no negative power to cancel out the positive power over the cycle. Electrical equipment possessing resistance as a single circuit component are electric fires, storage heaters and water heaters, all of which are designed to consume power and thereby create heat energy.

Inductance

An **inductor** is a coil or winding which possesses both resistance (R) and inductance (L). The unit of inductance is the **henry** (H). Figure 2.5 shows an inductor connected to an a.c. supply. Its opposition to current flow is called **impedance** (Z) which is measured in ohms. In order to express inductance in ohms you have to find the inductor's **inductive reactance** (X_L). This is given by the formula:

$$X_L = 2\pi fL \qquad [2.5]$$

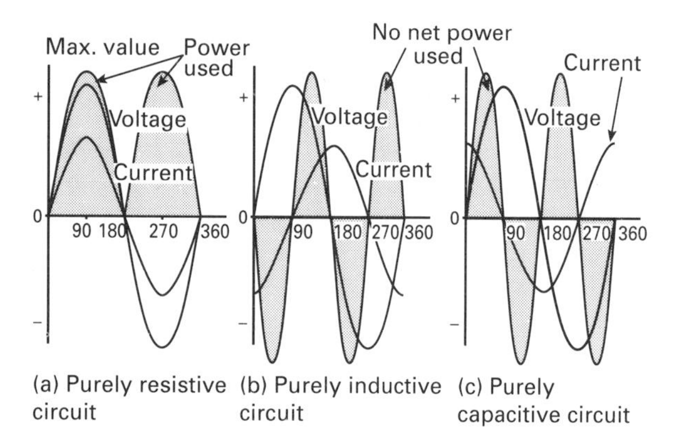

Figure 2.4 Power curves for resistive, inductive and capacitive circuits

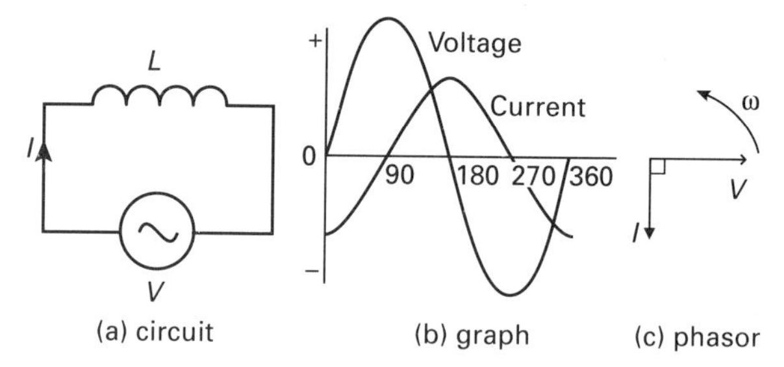

Figure 2.5 Circuit, graph and phasor diagram for 'purely' inductive a.c. circuit

Once the reactance is found the impedance of the circuit can be found from the formula:

$$Z = \sqrt{R + X_L} \qquad [2.6]$$

You could also find the impedance with an ammeter and voltmeter and use the formula:

$$Z = V/I \qquad [2.7]$$

If the inductor in Figure 2.5 actually had no resistance, often described as a 'pure' inductor, its current quantity would be displaced from the voltage quantity by a phase angle of 90°. This phase 'lag' of the current is caused by the inductor's induced voltage as a result of its magnetic field cutting through its own windings. This voltage opposes the supply voltage which creates the current lag condition. The circuit is said to have a **lagging power factor**.

If the instantaneous values of voltage and current are again multiplied together over one complete cycle, no power will be consumed since positive and negative quarter cycles of power cancel out (see Figure 2.4(b).

When a coil is designed to be highly inductive, it possess negligible resistance and has an extremely low power factor. This implies that it will consume very little power. Its application in a.c. circuits is based on its property to produce artifical magnetism often to create movement, voltage changes or current limitation.

Capacitance

A capacitor is a component which possesses capacitance (C) as its chief property. Its unit is the **farad** (F). When connected to an a.c. supply, its plates are continually being charged and discharged owing to the positive and negative cycles. No current actually flows through the capacitor but charges are built up on its plates by electron flow before a voltage is established.

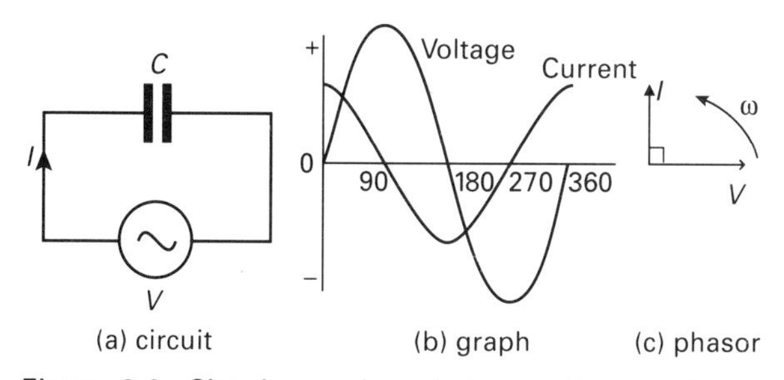

Figure 2.6 Circuit, graph and phasor diagram for capacitor connection in a.c. circuit

Figure 2.6 shows the voltage and current waveforms for a purely capacitive circuit and you will see that the current quantity ‘leads’ the voltage quantity by 90°. This implies that a capacitor causes a **leading power factor** and by multiplying together instantaneous voltages and currents, you will see from Figure 2.4(c) that the capacitor consumes no power. If a voltmeter and ammeter were connected in the capacitor’s circuit the ratio voltage/current would give capacitive reactance (X_C) and like inductive reactance is measured in ohms.

Capactive reactance can be found from the formula:

$$X_C = 1/2\pi fC \qquad [2.8]$$

RLC circuits

It is very important that you remember the three basic types of phasor diagram and also treat circuit symbols in their ‘pure’ state unless instructed otherwise. Moreover, since RLC circuits often show a mixture of components and are either connected in series or parallel, it would be advisable to apply the following steps when constructing phasor diagrams:

1) For series connected circuits the supply current flows through all components. Make the current a horizontal reference line, ending it as a closed arrow and marking it I_{ref}.
2) For parallel circuits, make the supply voltage the horizontal reference line since it is applied across all the connected circuit components (unless some components are in series with each other). Mark the line V_{ref} drawn as an open arrow.
3) Branch currents or potential differences should be drawn to a suitable scale and you must apply the current and voltage relationships already explained.
4) Using your protractor, you can find the phase angle between the supply voltage and supply current which should confirm any calculation.
5) It is important to label the phasor diagram with the proper letters and subscripts as well as show the direction of rotation. Remember that the current and voltage quantities are actually moving.

The following examples show how mixed RLC circuits can be solved.

Resistance and inductance in series

Q

Example 2.1

Figure 2.7 shows a circuit diagram of a resistor of 30 ohms connected in series with an inductor of negligible resistance having an inductive reactance of 40 ohms. If the supply to the circuit is 250 V and the frequency 50 Hz:

(i) determine
  a) impedance of the circuit
  b) current in the circuit
  c) pd across each component
  d) circuit power factor
  e) inductance of the coil;

(ii) construct a phasor diagram of the circuit.

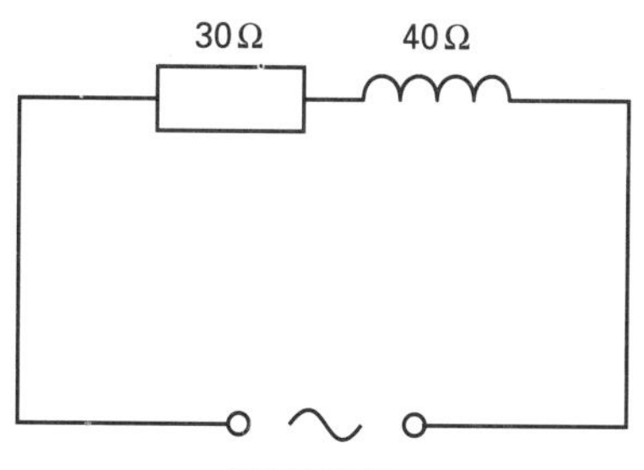

Figure 2.7 RL series circuit

A

Solution (i)

a) Since:

$$Z = \sqrt{R^2 + X^2}$$

then

$$Z = \sqrt{30^2 + 40^2}$$

$$= 50\ \Omega$$

b) Since:

$$I = V/Z$$

then

$$I = 250/50$$
$$= 5 \text{ A}$$

c) The pd's are as follows:

resistive part:

$$V = IR$$
$$= 5 \times 30$$
$$= 150 \text{ V}$$

the inductive part:

$$V = IX_L$$
$$= 5 \times 40$$
$$= 200 \text{ V}$$

You should note that the algebraic sum of the potential differences is 350 V yet the supply voltage is 250 V. This will be made clear when you study the phasor diagram.

d) Figure 2.8 shows how the components can be presented as an impedance triangle. The ratio of the sides R and Z can be used to find the power factor, i.e.

$$\cos \phi = R/Z$$
$$= 30/50$$
$$= 0.6 \text{ lagging}$$

It is lagging because of the coil's inductance.

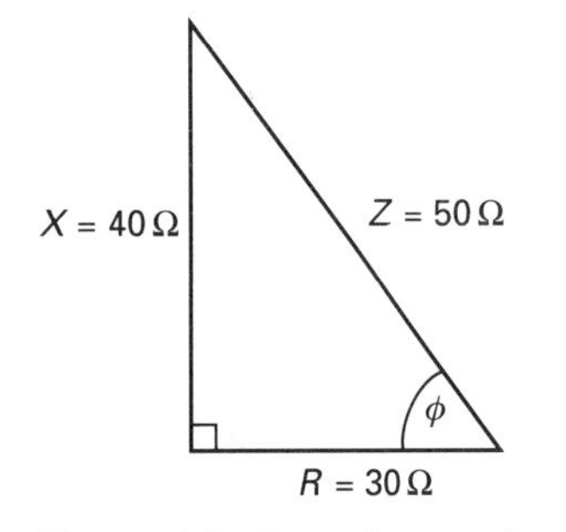

Figure 2.8 Impedance triangle

e) The inductance of the coil is found by using formula [2.5] and re-arranging it to find L:

Since

$$X_L = 2\pi fL$$
$$L = X_L/2\pi f$$
$$= 40/314.2$$
$$= 0.127 \text{ H}$$

A

Solution (ii)

The phasor diagram is shown in Figure 2.9. Note that the supply current is in phase with the pd across the resistance component and the pd across the inductive component is 90° out-of-phase with the reference current. By constructing a parallelogram of these two pd's the supply voltage is found to be the diagonal line – it

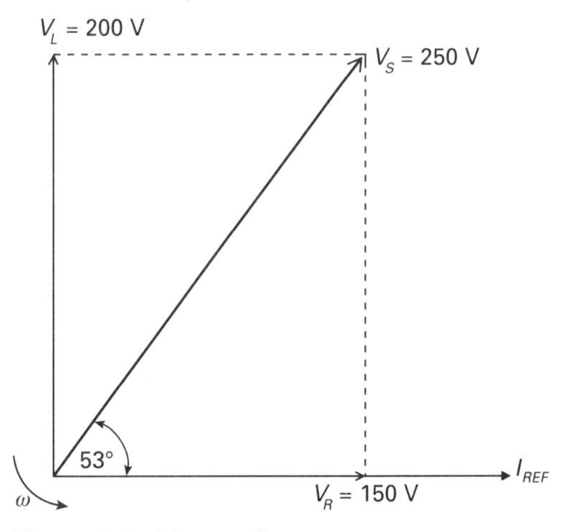

Figure 2.9 Phasor diagram

is the phasor sum of the pd's which can be found using the formula

$$V_S = \sqrt{V_R^2 + V_L^2}$$

Q

Example 2.2

Figure 2.10 shows an inductive coil connected to an a.c. supply of 240 V 50 Hz. An ammeter in the circuit reads 2 A. If the coil is connected to a d.c. supply of 100 V and

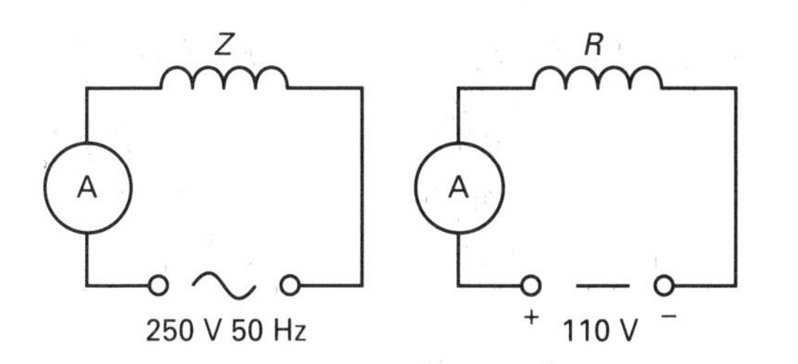

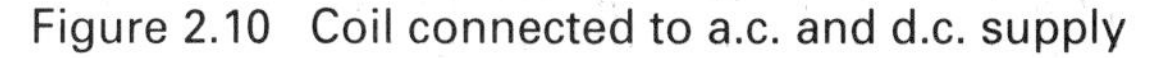
Figure 2.10 Coil connected to a.c. and d.c. supply

the ammeter reads 10 A, what is the coil's inductive reactance and inductance? Determine the power factor of the circuit.

A

Solution

On a.c.

$$Z = V/I = 240/2 = 120\ \Omega$$

On d.c.

$$R = V/I = 100/10 = 10\ \Omega$$

Re-arranging formula [2.6]:

$$X_L = \sqrt{Z^2 - R^2}$$

$$= \sqrt{120^2 - 10^2}$$

$$= 119.6\ \Omega$$

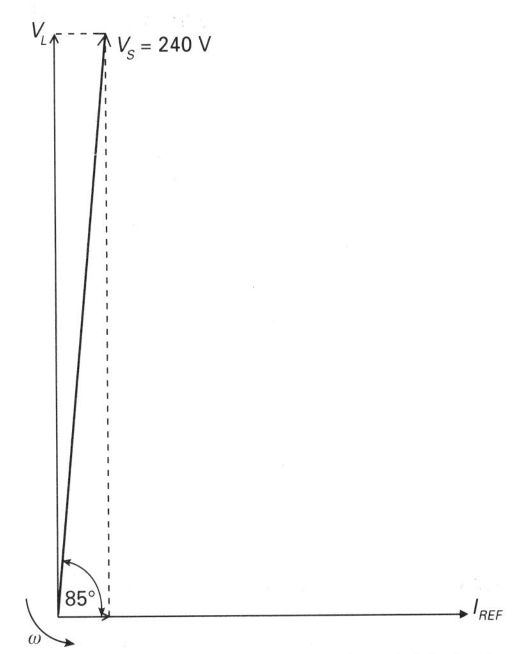

Figure 2.11 Phasor diagram for highly inductive circuit

Re-arranging formula [2.5]:

$$L = X_L/2\pi f$$

$$= 119.6/314.2$$

$$= 0.38\ \text{H}$$

The coil is very inductive and its power factor is again found by using the formula:

$$p.f. = R/Z = 10/120$$

$$= 0.08\ \text{lagging}$$

You will see that this is a very poor power factor causing the current to lag behind the supply voltage by an phase angle of 85°. A phasor diagram is shown in Figure 2.11.

Resistance and capacitance in series

Example 2.3

Figure 2.12 shows a 50 μF capacitor connected in series with a 60 Ω non-inductive resistor across a 240 V, 50 Hz supply. Determine:

a) the capactive reactance of the capacitor
b) the impedance of the circuit
c) the current consumed by the circuit
d) the power factor and phase angle of the circuit
e) the power consumed

Draw a phasor diagram showing the current and voltage relationship.

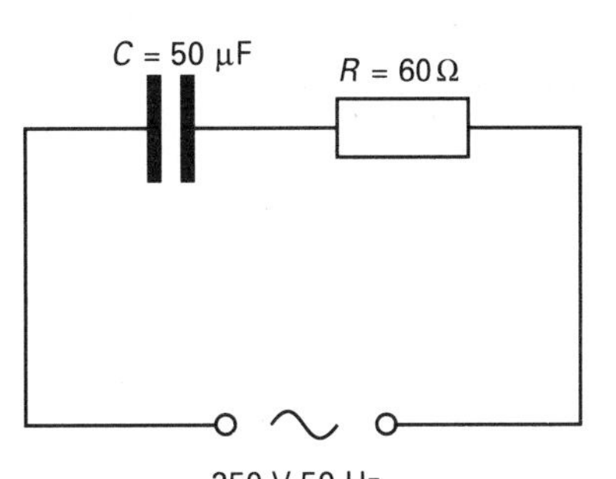

Figure 2.12 RC series circuit

Solution

a) From formula [2.8]

$$X_C = 1/2\pi fC$$

Since C is in μF then

$$X_C = 10^6/(314.2 \times 50)$$
$$= 63.65\ \Omega$$

b) The impedance

$$Z = \sqrt{R^2 + X^2}$$
$$= \sqrt{60^2 + 63.65^2}$$
$$= 87.47\ \Omega$$

c) The circuit current

$$I = V/Z$$
$$= 240/87.47$$
$$= 2.74\ \text{A}$$

d) The power factor

$$\cos\phi = R/Z$$
$$= 60/87.47$$
$$= 0.68 \text{ leading}$$

From this the phase angle $\phi = 46.69°$

e) The power consumed is given by formula [2.3]:

$$P = VI\cos\phi$$
$$= 240 \times 2.74 \times 0.68$$
$$= 447.2\ \text{W}$$

The phasor diagram is shown in Figure 2.13. You should note the pd across the resistor is found from $V_R = IR$ and the pd across the capacitor is found from $V_C = IX_C$. You should also note that if V_R was a higher value then the phase angle of the circuit would decrease.

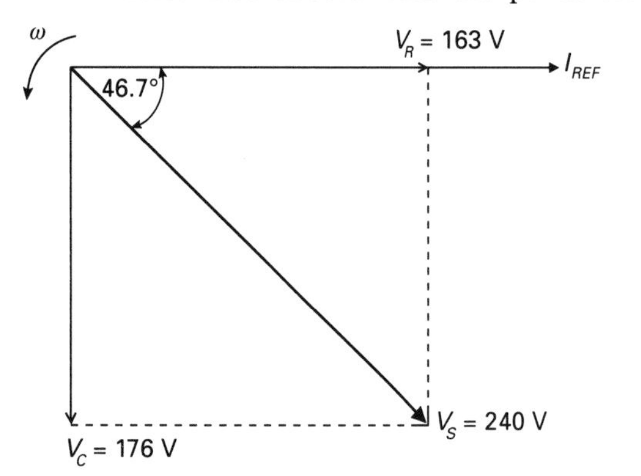

Figure 2.13 Phasor diagram for RC series circuit

Resistance and inductance in parallel

Example 2.4

Figure 2.14 shows a non-inductive resistor of 40 Ω is connected in parallel with a coil of inductance 95.6 mH and negligible resistance. If the supply voltage is 240 V, 50 Hz, determine:

a) the inductive reactance of the coil
b) the current through each circuit component
c) the supply current, power factor and phase angle.

Draw a phasor diagram of the circuit using a suitable scale.

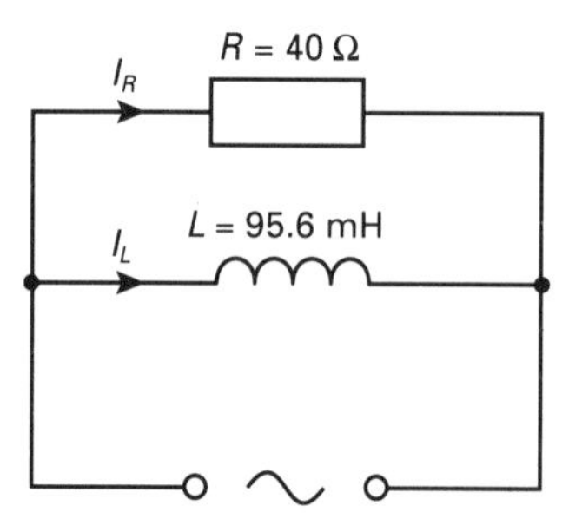

Figure 2.14 RL parallel circuit

Solution

a) $$X_L = 2\pi fL$$
$$= 314.2 \times 0.0956$$
$$= 30\ \Omega$$

b) $$I_R = V_S/R = 240/40$$
$$= 6\ \text{A}$$
$$I_L = V_S/X_L = 240/30$$
$$= 8\ \text{A}$$

c) The supply current is solved by using

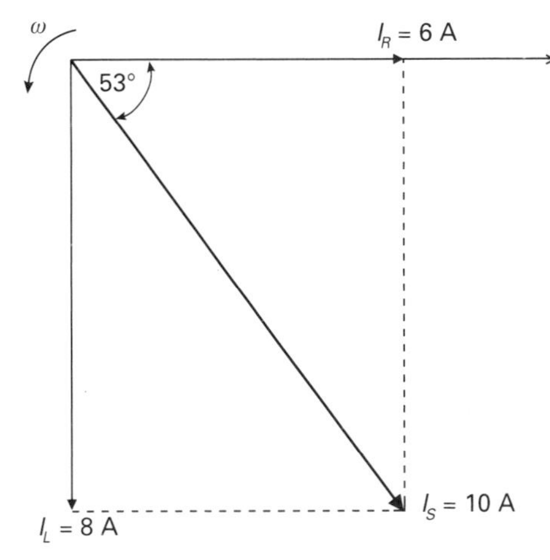

Figure 2.15 Phasor diagram for RL parallel circuit

Pythagoras' theorem (see Figure 2.15.) It is given by the formula:

$$I_S = \sqrt{I_R^2 + I_L^2}$$

The ratio of sides I_R and I_S can be used to find the power factor:

$$I_S = \sqrt{6^2 + 8^2} = 10 \text{ A}$$

Power factor

$$\cos \phi = I_R/I_S = 6/10$$
$$= 0.6 \text{ lagging}$$

Phase angle

$$\phi = 53°$$

Resistance and capacitance in parallel

Example 2.5

Figure 2.16 shows a non-inductive resistor of 20 Ω is connected in parallel with a capacitor having a capacitive reactance of 20 Ω. If the supply voltage is 240 V, 50 Hz determine:

a) the capacitor's capacitance
b) the current through each component
c) the supply current
d) power factor and
e) phase angle

Draw a phasor diagram of the circuit using a suitable scale.

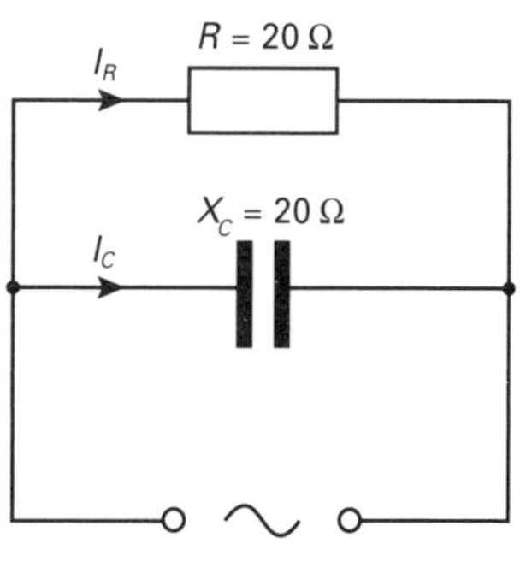

Figure 2.16 RC parallel circuit

A

Solution

a) $C = 1/2\pi f X_C$
$= 10^6/(314.2 \times 20)$
$= 159\ \mu\text{F}$

b) $I_R = V_S/R = 240/20$
$= 12 \text{ A}$
$I_C = V_S/X_C = 240/20$
$= 12 \text{ A}$

c) $I_S = \sqrt{I_R^2 + I_C^2}$
$= \sqrt{12^2 + 12^2}$
$= 16.97 \text{ A}$

d) $\cos \phi = I_R/I_S$
$= 12/16.97 = 0.707$
$= 45°$

The phasor diagram is shown in Figure 2.17. As previously mentioned, it is often an inductor which

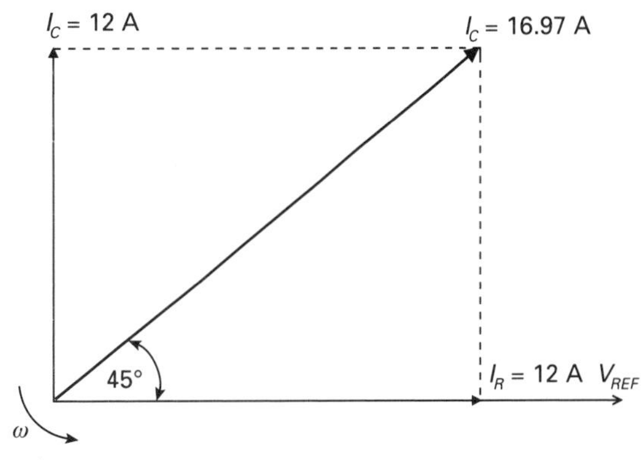

Figure 2.17 Phasor diagram for RC parallel circuit

needs power factor correction and this is achieved by placing a capacitor in parallel with it. Discharge lamp control gear is a typical application of such an arrangement.

The following example shows how the supply current can be reduced by 50% with the inclusion of a capacitor in circuit.

Q

Example 2.6

Figure 2.18 shows an inductive coil in parallel with a capacitor. If the inductor takes a current of 6 A and lags behind the supply voltage by a phase angle of 45°, what will be the supply current if the capacitor's current is 3 A and leads the supply voltage by 90°?

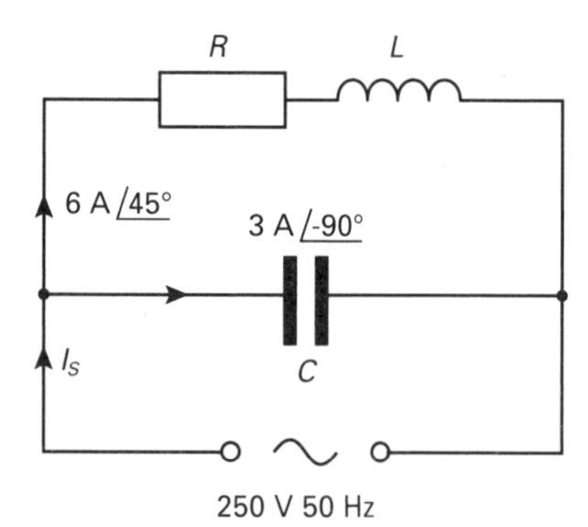

Figure 2.18 Power factor correction

A

Solution

The phasor diagram is shown in Figure 2.19. You should construct your diagram along the following lines:

1) Draw a horizontal line and label it the reference voltage;
2) Choose a suitable scale (say 1 cm = 1 A) and with your ruler and protractor, draw the inductor's current I_L of 6 A lagging the supply voltage by 45°;
3) Now draw the capacitor's current I_C of 3 A leading the supply voltage by 90°;
4) Construct a parallelogram and insert the diagonal line representing the supply surrent I_S which should measure 4.4 cm;
5) Use your protractor to verify the angle of 16°;

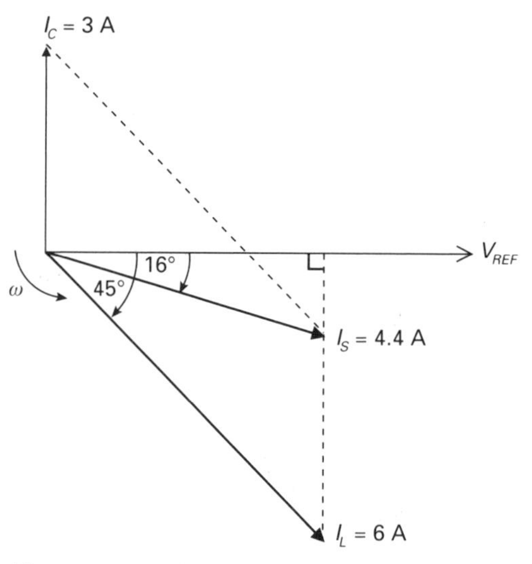

Figure 2.19 Effect of capacitor

6) On your calculator you will find that cos 45° = 0.707 and cos 16° = 0.96 which shows how much the circuit power factor has been improved by the addition of a capacitor. The supply current is reduced by 1.6 A.

It should become a little clearer to you that inductors and capacitors have the effect of neutralizing each other when connected in the same circuit. The next example shows all three components connected in series with each other. This also serves to illustrate the anti-phase effect of two dissimilar reactances which cause energy to flow backwards and forwards giving rise to a high voltage across each component's terminals. You will come across the term **resonance** which is often used to describe this condition of energy oscillation.

Example 2.7

Figure 2.20 shows a series circuit comprising a resistor of 5 Ω, an inductor of 0.02 H and a capacitor of 150 μF. The components are connected to a single-phase a.c. supply of 240 V, 50 Hz. Determine:

a) the resultant reactance;
b) the impedance of the circuit;
c) the supply current;
d) the circuit power factor;
e) the p.d. across each component;
f) the frequency at which $X_L = X_C$

Draw a phasor diagram of the circuit.

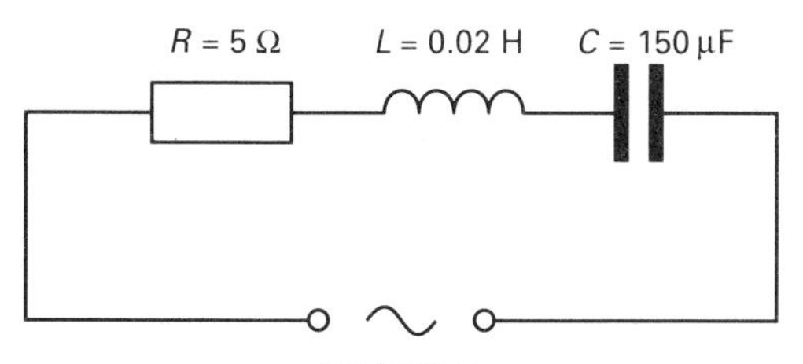

Figure 2.20 RLC series circuit

A

Solution

From formula [2.5]:

$$X_L = 2\pi fL = 314.2 \times 0.02$$
$$= 6.28\ \Omega$$

From formula [2.8]:

$$X_C = 1/2\pi fC = 10^6/(314.2 \times 150)$$
$$= 21.2\ \Omega$$

This tells you that the circuit condition favours a leading power factor condition since X_C has a higher value than X_L. Since the reactances are in anti-phase with each other (180° apart), they must be subtracted.

a) resultant reactance

$$X = X_C - X_L = 21.2 - 6.28$$
$$= 14.92\ \Omega$$

b) impedance

$$Z = \sqrt{R^2 + X^2}$$
$$= \sqrt{5^2 + 14.92^2}$$
$$= 15.75\ \Omega$$

c) supply current

$$I = V/Z = 240/15.75$$
$$= 15.2\ \text{A}$$

d) power factor

$$\cos\phi = R/Z = 5/15.75$$
$$= 0.32 \text{ leading}$$

e) pd across R

$$V_R = IR = 15.2 \times 5$$
$$= 76\ \text{V}$$

pd across X_L

$$V_L = IX_L = 15.2 \times 6.28$$
$$= 103.36\ \text{V}$$

pd across X_C

$$V_C = IX_C = 15.2 \times 21.2$$
$$= 322.24\ \text{V}$$

f) **Resonant frequency** (f_r) occurs when $X_L = X_C$. Since $X_L = 2\pi fL$ and $X_C = 1/2\pi fC$ then by transposition of the formula:

$$f_r = 1/(2\pi\sqrt{LC})$$
$$= 1/(2\pi \times \sqrt{0.02 \times 150 \times 10^{-6}})$$
$$= 92\ \text{Hz}$$

Figure 2.21 is a phasor diagram of the circuit.

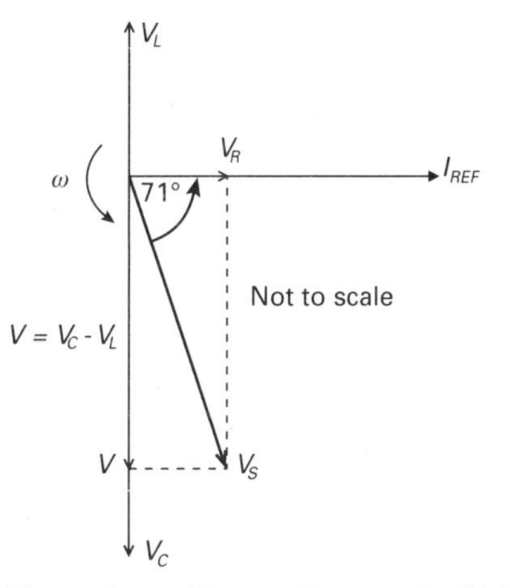

Figure 2.21 Phasor diagram for RLC series circuit

Power factor

Power factor was previously explained as the relationship between current and voltage for three types of RLC component. Several worked examples made reference to the impedance triangle as a method of obtaining the cosine of the phase angle which expressed power factor as a value between 0 and 1. Other right-angled triangles could be used, for example, Figure 2.22 (taken from Figure 2.15) shows the supply current and

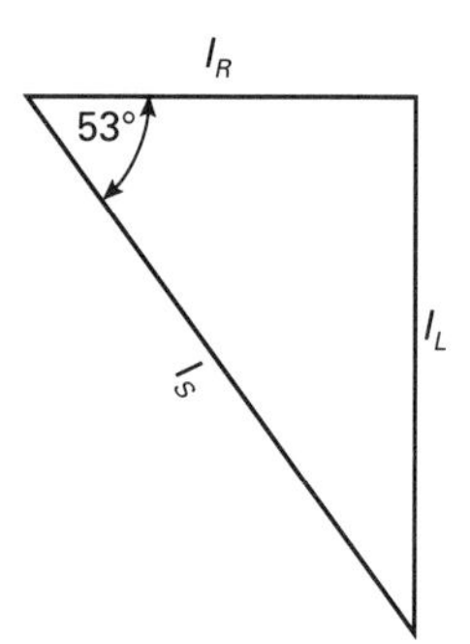

Figure 2.22 Current triangle for a parallel circuit

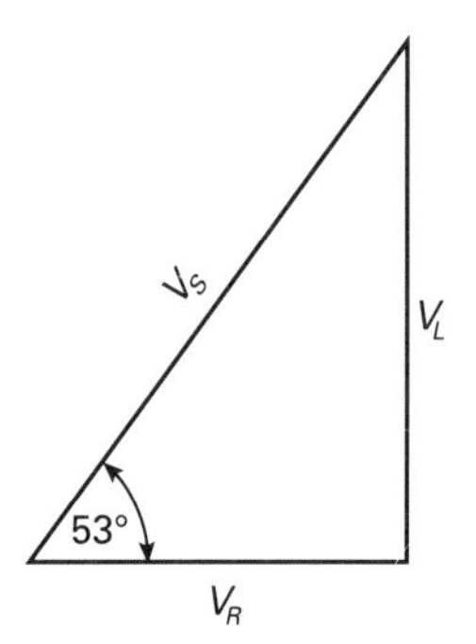

Figure 2.23 Voltage triangle for a series circuit

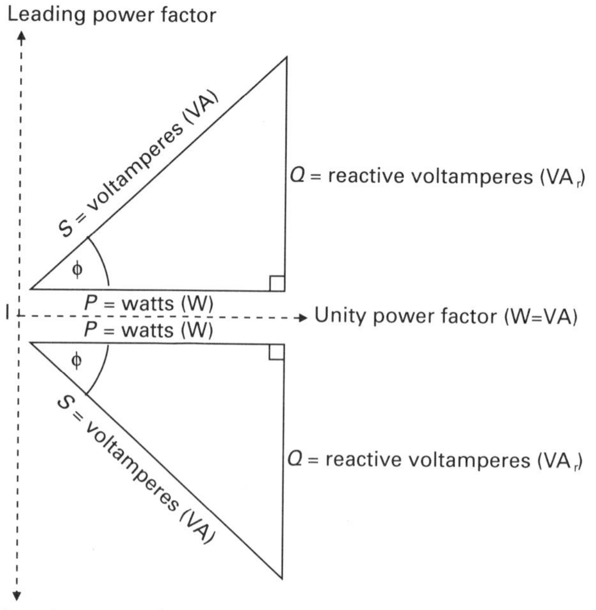

Figure 2.24 Power triangles showing power factor conditions

branch currents for parallel connected components and Figure 2.23 (taken from Figure 2.9) shows the supply voltage and potential differences for series connected components. It is also possible to find power factor using a **power triangle** since it was shown in formula [2.4] that:

$$p.f. = P/VI$$

The quantity (P) represents the true power taken by the circuit as measured by a wattmeter. The quantity (VI) represents the apparent power or **voltamperes** (S) as measured by a voltmeter and ammeter. These quantities are shown in the power triangle of Figure 2.24. The line at 90° is the quadrature component or no-power quantity and refers to the extreme phase condition found in a 'pure' inductor or capacitor. This quantity is better known as the **reactive power** (VI_R) or **reactive voltamperes** (Q).

Power triangles truly represent power factor and are and ideal way of representing single-phase and three-phase circuits especially where quantity units are expressed in kilowatts (kW), kilovoltamperes (kVA) and reactive kilovoltamperes (kVAr). You should note that if a circuit is inductive the right-angled triangle is drawn downwards showing a lagging condition.

If a **power factor improvement capacitor** is inserted in such a circuit, the line marked Q will decrease in length as a result of the phase angle decreasing and the only way to make the lagging Q disappear completely is to insert in the circuit a capacitor of the correct value which will result in unity power factor. In other words, inject a leading Q line of the same magnitude which can be subtracted from the lagging Q line leaving no phase angle between P and S.

The following examples will help to show how these problems can be solved.

Example 2.8

The circuit connections of a single-phase a.c. induction motor are shown in Figure 2.25. The wattmeter reads 5 kW, voltmeter reads 240 V and ammeter reads 32 A.

(a) Determine the power factor of the circuit.

(b) Draw a power triangle and determine graphically or by calculation the reactive voltamperes in kVAr.

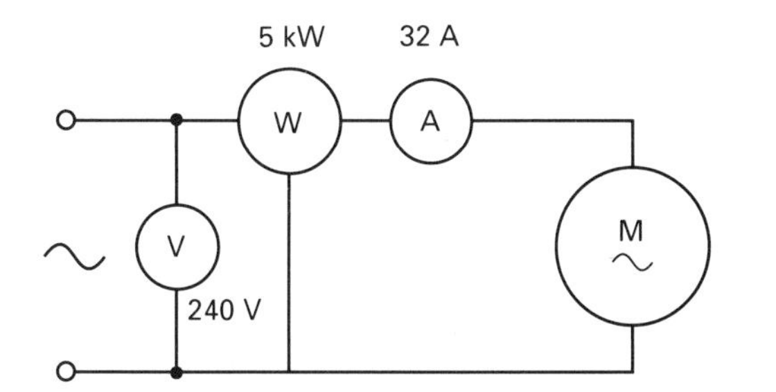

Figure 2.25 Instrument connections for a motor circuit

Solution

(a) Power factor

$$\text{p.f.} = P/S$$
$$= 5000/(240 \times 32)$$
$$= 0.65 \text{ lagging}$$

(b) Figure 2.26 shows the power triangle.

$$P = 5 \text{ kW}$$
$$S = 240 \times 32$$
$$= 7.68 \text{ kVA}$$

The lagging kVAr is found by re-arranging Pythagoras' theorem in the formula:

$$S = \sqrt{P^2 + Q^2}$$

then

$$Q = \sqrt{S^2 - P^2}$$
$$= \sqrt{7.68^2 - 5^2}$$
$$= 5.83 \text{ kVAr}$$

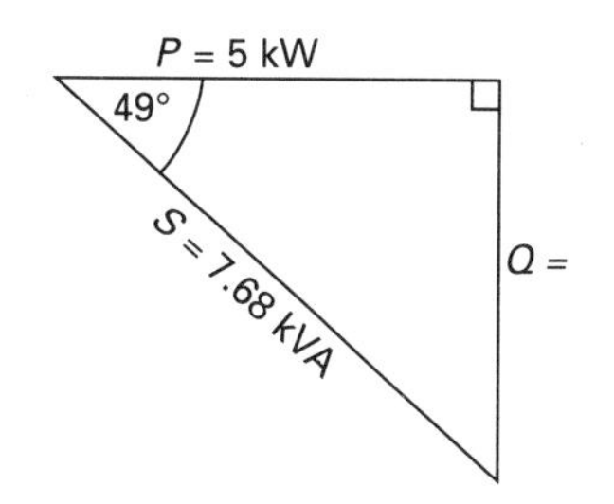

Figure 2.26 Power triangle

Example 2.9

A 240 V, 50 Hz single-phase induction motor has a power factor of 0.75 lagging and takes a supply current of 39 A. Determine the following:

a) input power (kW);
b) apparent power (kVA);
c) capacitor size to improve p.f. to unity.

Solution

a) Since

$$P = VI \cos \phi$$
$$= 240 \times 39 \times 0.75$$
$$= 7020 \text{ W}$$
$$= 7.02 \text{ kW}$$

b) Since

$$S = VI$$
$$= 240 \times 39$$
$$= 9360 \text{ VA}$$
$$= 9.36 \text{ kVA}$$

c) The lagging kVAr is found by re-arranging Pythagoras' theorem as shown above.

Hence

$$Q = \sqrt{S^2 - P^2}$$
$$= \sqrt{9.36^2 - 7.02^2}$$
$$= 6.19 \text{ kVAr}$$

To raise the power factor to unity in the above example the kVAr or (Q) line shown in Figure 2.24 must be cancelled out by an equal and opposite leading kVAr, i.e. it too must be 6.19 kVAr. A

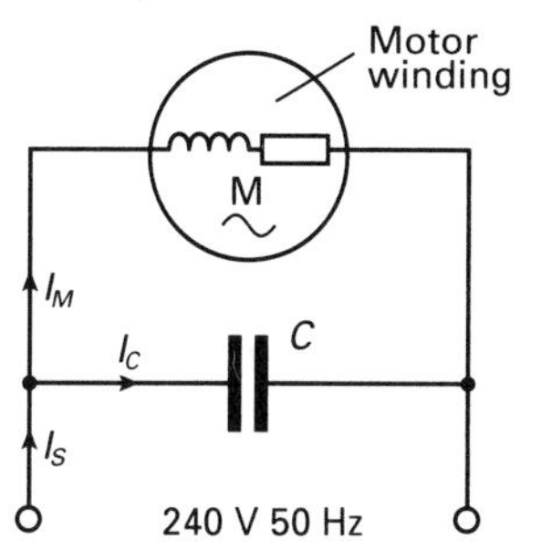

Figure 2.27 Motor circuit

diagram of the motor connections is shown in Figure 2.27. Notice that the capacitor is connected across the 240 V supply terminals.

The current (I_C) is found by the formula

$$I_C = VA_R/V$$
$$= 6194/240$$
$$= 25.8 \text{ A}$$

To find the capacitor's value you must firstly find its capacitive reactance which is expressed as:

$$X_C = V/I_C$$
$$= 240/25.8$$
$$= 9.3 \ \Omega$$

By re-arranging formula [2.8] the capacitance is:

$$C = 1/(2\pi f X_C)$$

Expressed in microfarads:

$$C = 10^6 / (2 \times \pi \times 50 \times 9.3)$$
$$= 342 \ \mu\text{F}$$

You should note that by reducing the power factor to unity the motor only takes 29.25 A instead of 39 A. Figure 2.28 shows a phasor diagram of the currents.

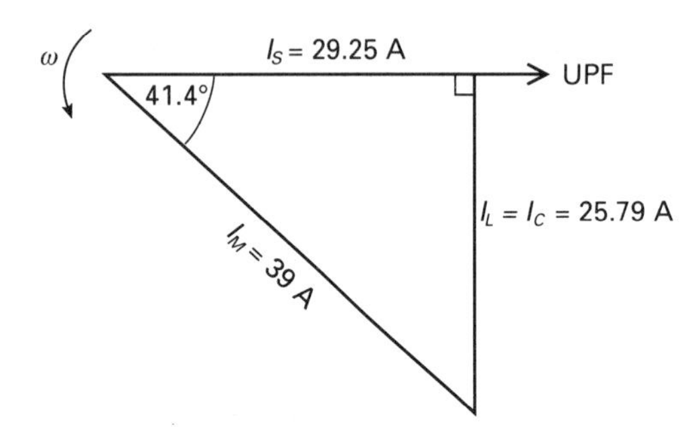

Figure 2.28 Phasor diagram

Example 2.10

A 50 kW a.c. motor is 83% efficient and operates at a power factor of 0.65 lagging. If a 40 kVAr power factor improvement capacitor is to be connected in parallel with the motor, find graphically or by calculation the input kVA before and after the capacitor is connected.

Solution

Since efficiency is the ratio output/input, then the motor's input can be expressed as:

$$\text{Input power } (P_I) = \frac{\text{Output } (P_o)}{\text{per unit efficiency}}$$
$$= 50/0.83 = 60.24 \text{ kW}$$

Since the input power is electrical, from formula [2.4]:

$$p.f. = \frac{P}{VI} = \frac{P}{S}$$

therefore

$$S = P/p.f.$$
$$= 60.24/0.65 = 92.67 \text{ kVA}$$

The motor's lagging reactive voltamperes (Q) is found from the formula:

$$Q = \sqrt{S^2 - P^2}$$
$$= \sqrt{92.67^2 - 60.24^2}$$
$$= 70.43 \text{ kVAr}$$

This kVAr is reduced by the capacitor's leading kVAr, hence resultant kVAr is:

$$Q = \sqrt{70.43^2 - 40^2} = 30.43 \text{ kVAr}$$

When the motor is running corrected the kVA is found to be:

$$S = \sqrt{P^2 + Q^2}$$
$$= \sqrt{60.24^2 + 30.43^2}$$
$$= 67.49 \text{ kVA}$$

In the above example you should note the power factor has improved to 0.89 (see Figure 2.29)

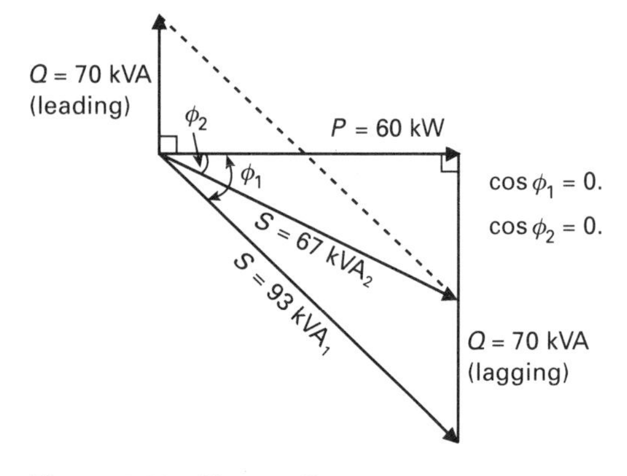

Figure 2.29 Phasor diagram

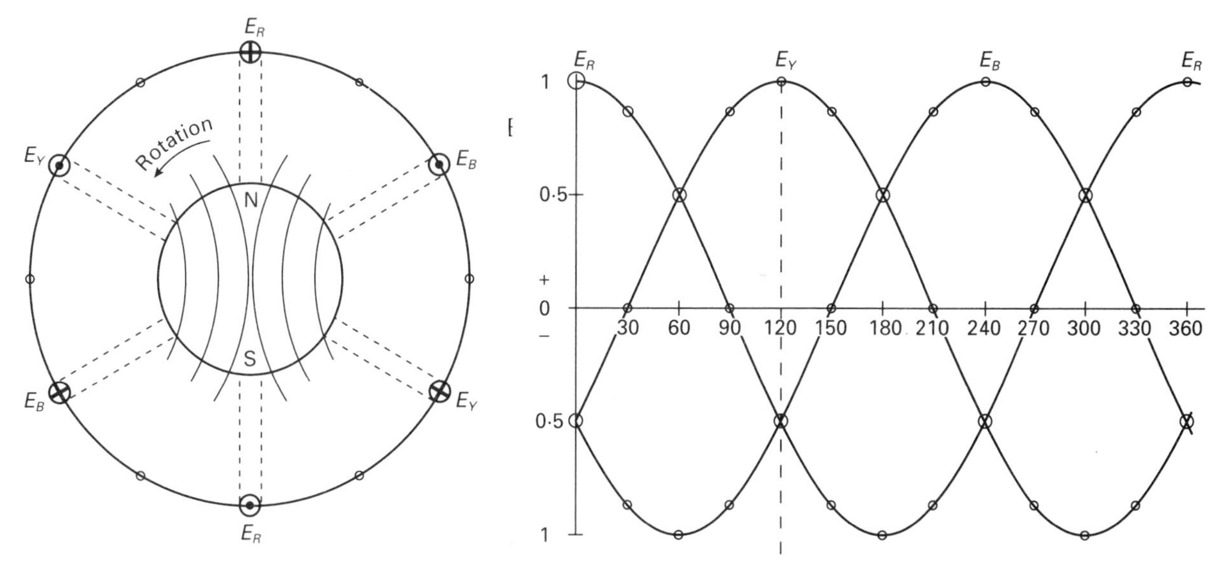

Figure 2.30 Three-phase generation of a.c.

Three-phase supplies

Figure 2.30 shows how three phases are spaced at 120° intervals inside an a.c. generator. The conductor phases (in pairs) are identified as red, yellow and blue and when the generator's rotating magnetic field sweeps across these phases they each produce a voltage in the sequence shown. You will see that when one phase reaches a maximum in any direction the other two phases reach only half their maximum value. This means, for example, that if the yellow phase was delivering an instantaneous current of say 30 A in the positive direction, then the red and blue phases would each return an instantaneous current of 15 A in the negative direction to be in balance.

At distribution level to consumers, the three-phase system would be given a neutral reference conductor since there is no guarantee that the system could remain balanced. Any of the phases may take different load currents to that of other phases and the purpose of a neutral conductor is to allow any out-of-balance current to flow back through the system (see Figure 2.31).

In dealing with three-phase systems you will come across connections known as **star** and **delta**. The term √3 is used to distinguish between line and phase values of the current and voltage. To illustrate this, Figure 2.32(a) shows a typical star-connected system. For clarity reasons, only the red phase voltage (E_{NR}) and red to yellow line voltage (E_{YNR}) are dealt with. The phasor diagram shows the induced voltages for each of the phases and these are labelled from the neutral or star point outwards. You will see that the yellow phase to neutral voltage E_{NY}, when reversed becomes E_{YN}. This phase voltage is vectorially added to E_{NR} and the resultant called E_{YNR}. This is the line value between both red and yellow phases. By measurement it is found to be √3 (1.732) longer than the phase value. You should complete Figure 2.32(b) for the other two line values.

The following example shows the importance of √3 in star and delta systems, making clear the difference between line and phase values for each system.

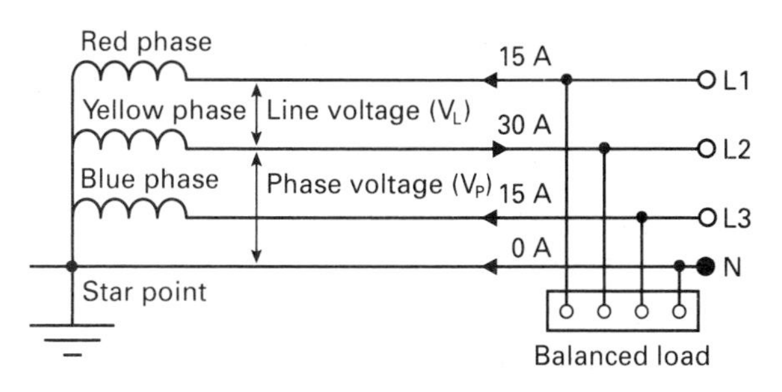

Figure 2.31 Neutral connection for keeping supply. system balanced

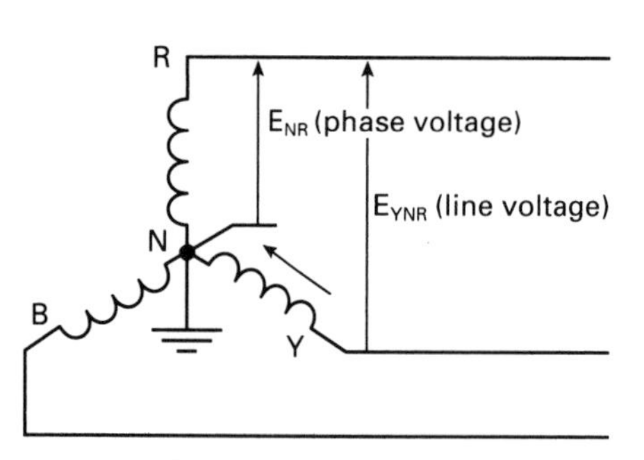

(a) Phase and line voltage

(b) Phasor diagram

Figure 2.32 Star connnected system

Example 2.11

Figure 2.33 represents a three-phase star connected supply system feeding a delta connected load. If the star connected phase voltage is 240 V and its phase current is 20 A, determine the star line voltage and current and also the delta line and phase values of voltage and current.

Solution

In star

$$I_L = I_P$$

therefore

$$I_L = 20 \text{ A}$$

Also $V_L = \sqrt{3}\, V_P$

therefore

$$V_L = 3 \times 240$$

$$= 415 \text{ V}$$

In delta

$$I_L = \sqrt{3}\, I_P$$

therefore

$$I_P = 20/3$$

$$= 11.56 \text{ A}$$

Also $V_L = V_P$

therefore

$$V_L = 415 \text{ V}$$

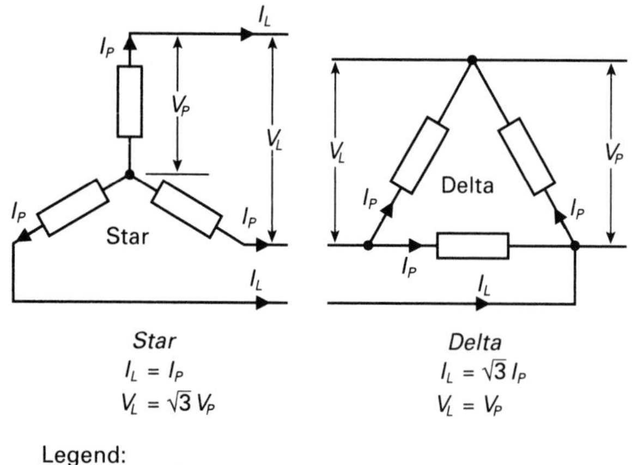

Figure 2.33 Star and delta connections

When dealing with three-phase system the total power is given by:

$$P = \sqrt{3}\, V_L I_L \cos \phi \qquad [2.9]$$

The power in each phase is given by:

$$P = V_P I_P \cos \phi \qquad [2.10]$$

The following are some examples of finding power in three-phase systems:

Example 2.12

Calculate the line current and total power consumed by three 40 ohm resistors connected in star to a 415 V three-phase star supply system.

Solution

The phase voltage across the load is:

$$V_P = V_L/\sqrt{3}$$
$$= 415/1.732$$
$$= 240 \text{ V}$$

The line current is the same as the phase current.

But

$$I_P = V_P/R$$
$$= 240/40$$
$$= 6 \text{ A}$$

Since the resistors are of equal value, this will be the current in all three lines and you should note that unity power factor exists. From the formula [2.9]:

$$P = \sqrt{3}\, V_L I_L \cos\phi$$
$$= 1.732 \times 415 \times 6 \times 1$$
$$= 4313 \text{ W} = 4.313 \text{ kW}$$

Example 2.13

Three impedances each having a value of 40 ohm and power factor 0.8 lagging are to be connected in (a) star and (b) delta across a 415 V three-phase supply. Determine the supply current and the total power consumed in each case.

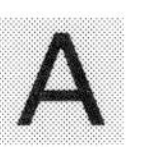

Solution

(a) In star

$$I = I_P = V_P/Z$$
$$= 240/40 = 6 \text{ A}$$
$$P = 3 \times 240 \times 6 \times 0.8$$
$$= 3456 \text{ W}$$

(b) In delta

$$I_P = 415/40 = 10.375 \text{ A}$$

and

$$I_L = \sqrt{3}\, I_P$$
$$= \sqrt{3} \times 10.375 = 17.97 \text{ A}$$
$$P = \sqrt{3} \times 415 \times 18 \times 0.8$$
$$= 10\,333 \text{ W}$$

Example 2.14

Figure 2.34 shows three consumers connected to a 415 V three-phase, four-wire system. If Consumer A has unity power factor, Consumer B has a power factor of 0.6 lagging and Consumer C has a power of 0.7 leading, determine the system's overall kW, kVA, kVAr and power factor.

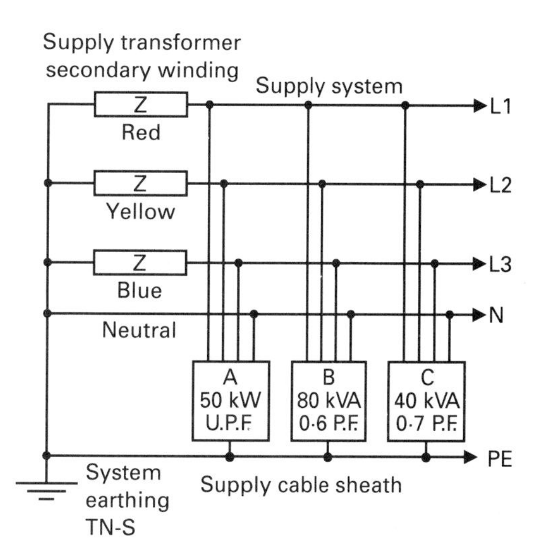

Figure 2.34 Three-phase four wire supply

Solution

Consumer A $\quad S = P/\cos\phi$

$$= 50/1 = 50 \text{ kVA}$$
$$Q = 0 \text{ kVAr}$$

Consumer B $\quad P = S \times \cos\phi$

$$= 80 \times 0.6$$
$$= 48 \text{ kW}$$
$$Q = S \times \sin\phi$$
$$= 80 \times 0.8 = 64 \text{ kVAr}$$

Consumer C $P = S \times \cos \phi$

$$= 40 \times 0.7 = 28 \text{ kW}$$

$$Q = S \times \sin \phi$$

$$= 40 \times 0.714$$

$$= 28.57 \text{ kVAr}$$

The total kilowatts used by the three consumers is:

$$P = 50 + 48 + 28$$

$$= 126 \text{ kW}$$

The total kVAr is the difference between the quadrature components, thus:

$$Q = Q_B - Q_C$$

$$= 64 - 28.57$$

$$= 35.43 \text{ kVAr}$$

The total kVA can be found by Pythagoras' theorem:

$$S = \sqrt{P^2 + Q^2}$$

$$= \sqrt{126^2 + 35.43^2}$$

$$= 130.9 \text{ kVA}$$

The overall power factor:

$$\cos \phi = P/S$$

$$= 126/130.9$$

$$= 0.96 \text{ lagging}$$

The power diagram is shown in Figure 2.35

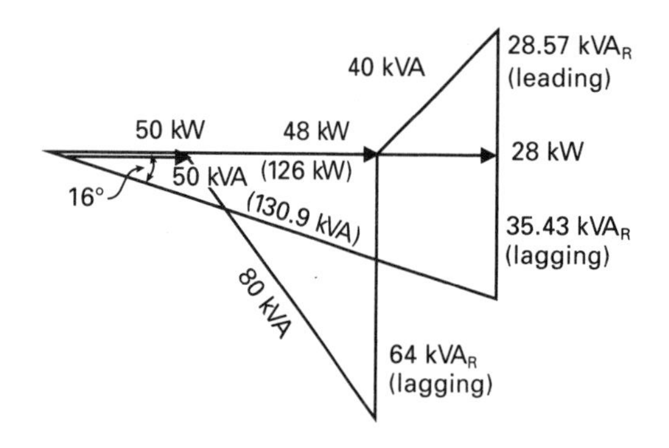

Figure 2.35 Graphical solution

Exercise 2

1. Figure 2.36 shows the circuit diagram of a SON discharge lamp. With switch (S) open, ammeter (A_1) reads 5 A and the wattmeter (W) reads 420 W. With the switch closed, ammeter (A_2) reads 2.2 A, ammeter (A_1) reads 3 A and the wattmeter reads 420 W. Draw a phasor diagram of the circuit using a scale of 1 A = 2 cm.

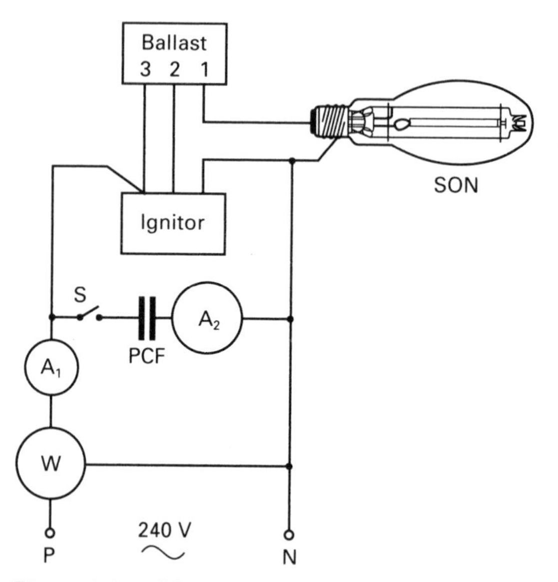

Figure 2.36 SON discharge lamp circuit

2. A 20 Ω resistor is connected in series with an inductor having an inductance of 0.2 H across a 240 V, 50 Hz supply. Ignoring the inductor's resistance:

   a) determine the following:

      (i) the inductive reactance of the inductor
      (ii) the impedance of the circuit
      (iii) the circuit current
      (iv) the potential difference across each component
      (v) the circuit power factor
      (vi) the power consumed.

   b) Draw a phasor diagram of the circuit.

3. A 20 μF capacitor is connected in series with a 50 Ω resistor across a 240 V, 50 Hz supply.

   a) Determine the following:

      (i) capacitive reactance of the capacitor

(ii) the circuit impedance
(iii) the circuit current
(iv) the potential difference across each component
(v) the circuit power factor
(vi) the power consumed.

b) Draw a phasor diagram of the circuit.

4. Figure 2.37 shows a series connected RLC circuit.

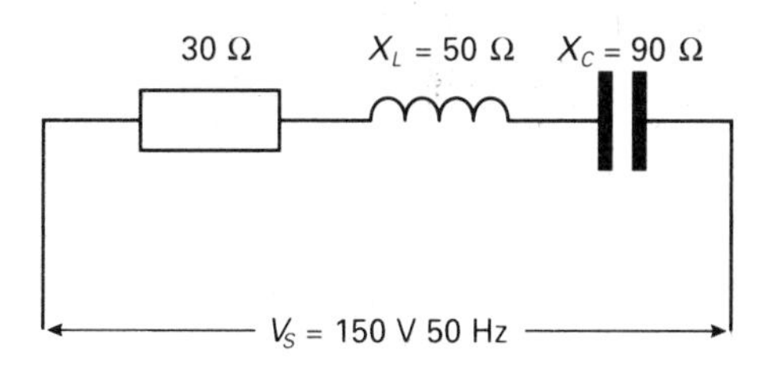

Figure 2.37 Series RLC circuit

a) Determine the following:

(i) impedance
(ii) circuit current
(iii) potential difference across each component
(iv) circuit power factor
(v) power consumed.

b) Draw a phasor diagram of the circuit.

5. With reference to Figure 2.38 determine the current in each ammeter and draw a phasor diagram of the circuit.

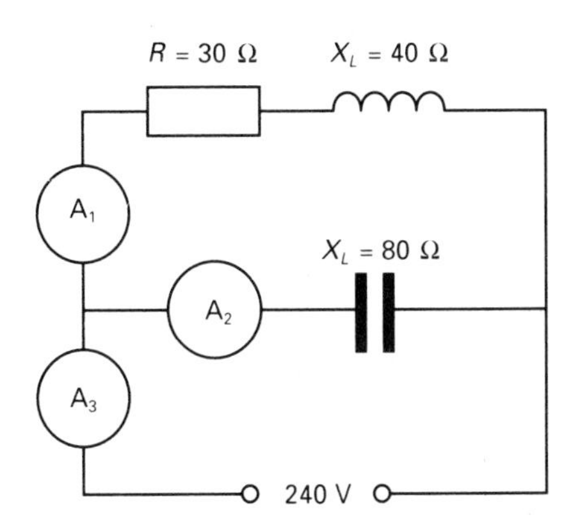

Figure 2.38 Mixed RLC circuit

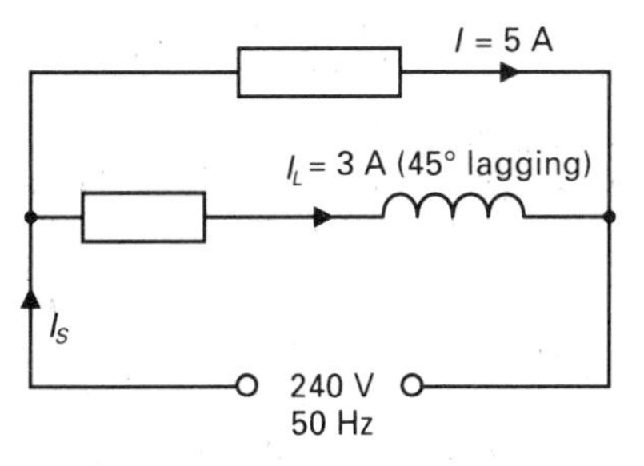

Figure 2.39 Series/parallel circuit

6. Figure 2.39 shows a series/parallel RL circuit.

a) Draw a phasor diagram of the circuit to a scale of 1 cm = 1 A

b) Determine the supply current and power factor of the circuit.

7. An a.c. current having a maximum value of 3 A lags behind an a.c. voltage having a maximum value of 300 V by a phase angle of 45°. Draw a graph of the quantities to suitable scale and determine their r.m.s. values.

8. A 12 kW/240 V single-phase load operates at a power factor of 0.7 lagging.

a) Calculate the current taken by the load.

b) Obtain graphically or by calculation the reactive voltamperes

9. A resistive heating load of 18 kW consists of three separate elements, each rated at 6 kW/240 V. The ends of the heating elements are wired separately into a six-terminal connection block:

a) Draw a circuit diagram of this arrangement, showing how the six terminals should be interconnected when fed from:

(i) a 415 V three-phase and neutral supply
(ii) a 240 V single-phase supply.

b) For both types of supply:

(i) calculate the current in the phase and neutral conductors

(ii) state the power consumed by the heating load if the neutral accidently became disconnected.

10. A factory has three-phase loads of (i) 120 kW unity power factor and (ii) 240 kVA at 0.8 power factor lagging. Draw a power triangle representing these two conditions and determine the total kW, kVA, kVAr and power factor of the system.

Electric motors and starters

Objectives

After reading this chapter you should be able to:

- *describe the principles of electromagnetic induction and conduction;*
- *perform calculations to find efficiency, torque and power of motors;*
- *describe the production of a rotating magnetic field from a three-phase supply;*
- *perform calculations to find synchronous speed, rotor speed and fractional slip;*
- *describe the operation and performance of three-phase cage rotor and wound rotor induction motors;*
- *describe the operation and performance of the following single-phase motors: resistance-start, capacitor-start, shaded-pole and universal;*
- *describe the function and operation of the following starters: direct-on-line, star-delta, autotransformer and rotor resistance;*
- *draw circuit diagrams of motors and starters as listed;*
- *describe the protective measures used in starters against overload and loss of supply voltage;*
- *state the reason for motor power factor improvement in motor circuits;*
- *state several methods that can be used to control the speed of motors.*

Basic Principles

When a current-carrying conductor is placed at right angles to a magnetic field a force (F) is exerted on it which tries to repel it out of the field. This repulsion is caused by the conductor's own magnetic field interacting with the primary magnetic field (see Figure 3.1). It is found that the magnitude of force on the conductor is proportional to:

1) the flux density (B) of the magnetic field measured in teslas (T);

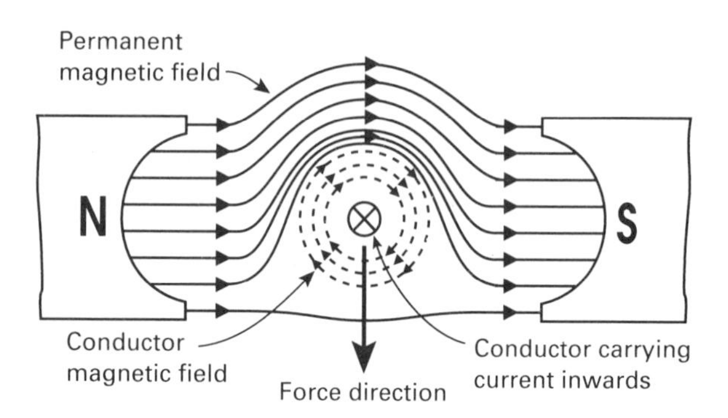

Figure 3.1 Force in a current-carrying conductor situated at right-angles to a magnetic field

2) the current (I) flowing in the conductor measured in amperes (A);
3) the effective length (l) of the conductor measured in metres (m).

Hence:

$$F = BIl \text{ newtons} \quad [3.1]$$

The interaction of two magnetic fields, causing the conductor to move, is the basic operating principle of most motors. The central idea is to have one of the magnetic fields on a motor's stationary part, known as the **frame** or **yoke** and the other magnetic field on the motor's rotating part. In an a.c. motor the rotating part is called a **rotor** whereas in a d.c. motor it is called an **armature**.

Electromagnetic induction (i.e. a moving field cutting conductors) is used to create the secondary magnetic field for a rotor-type motor whereas in an armature-type motor the secondary field is created by electromagnetic conduction (i.e. current fed into a commutator via brushgear). Both these methods will be discussed in this chapter.

Torque

The movement of a motor's rotating part is described by its torque (T) in newton metres which is the product of a tangential force (F) acting on its

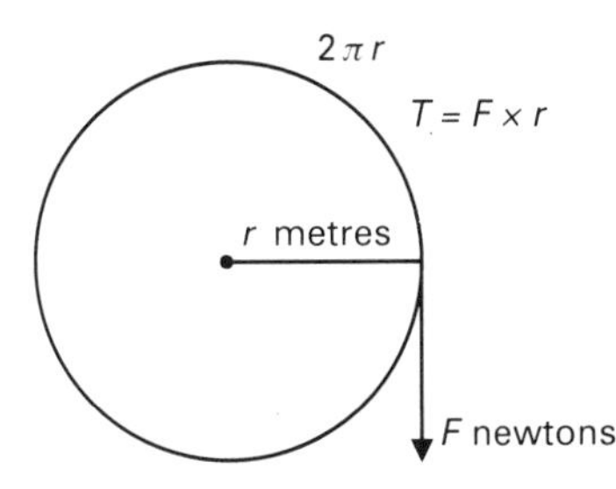

Figure 3.2 Force acting in a circular path on motor shaft

drive shaft or pully and the measured distance (r) from the centre of the shaft. This distance is the radius measurement (see Figure 3.2).
Hence:

$$T = Fr \text{ Nm} \qquad [3.2]$$

Torque is dependent upon the efficiency of the magnetic circuit and the air gap or clearance between the stationary and rotating parts of the motor. If the air gap was large it would create a weak magnetic circuit and this would result in a low torque. You will come across several meanings for the term torque, such as **starting torque** which is needed to overcome static friction when the motor is first switched on from standstill; **accelerating torque** which is needed to run the motor's rotor up to full speed; **running torque** which is the torque on the motor's shaft at full speed; **pull-out torque** which is the maximum torque for the motor to maintain its operating speed before it stalls.

Output power

A motor converts electrical energy into mechanical energy which is work done (W). If the motor's shaft makes one revolution, the force (F) on it will have moved through a distance of $2\pi r$ metres. Hence, work done is:

$$W = F \times 2\pi r \text{ joules} \qquad [3.3]$$

The motor's rate of working or output power (P_o) measured in watts (W) is the work done in one second. If its shaft rotates at n revolutions per second its mechanical output power will be:

$$P_o = F \times 2\pi r \times n$$

Since torque is force times radius (distance), i.e. $T = Fr$ (see expression [3.2], then $F = T/r$ and therefore:

$$P_o = T/r \times 2\pi r \times n$$

$$= 2\pi nT \text{ watts} \qquad [3.4]$$

Efficiency

Not all the electrical energy supplied to a motor is converted into mechanical energy. Between 20–30% is lost in various ways, mostly as heat energy. In the windings, resistive losses occur which are called **copper losses** and in the core, magnetising losses occur which are called **iron losses**. There are other losses such as **friction losses** from bearings and brushes and **windage losses** of rotating parts such as an integral fan to keep the motor cool.

As a general rule, motors having a high power output rating are more efficient than those having a low output power rating. Tests show that a motor's efficiency improves while it is on load. When left to idle on no-load it is likely to consume between 25–50% of its full-load running current. For this reason it is important to ensure that a motor does not spend a large proportion of its operating time on no-load or even on low-load. The rule of thumb in motor selection is to consult manufacturers of the connected driven machine and not to oversize a motor for its particular application.

Efficiency of a drive motor or its driven machine can be found from electrical input and mechanical output details. Since output cannot be greater than input, then percentage efficiency is given by the expression:

$$\% \text{ Efficiency } (\eta) = \frac{\text{output power} \times 100}{\text{input power}} \qquad [3.5]$$

Per unit efficiency is:

$$\text{Efficiency } (\eta) = \frac{\text{output power}}{\text{input power}} \qquad [3.6]$$

If consideration is made to losses, then:

$$\text{Efficiency } (\eta) = \frac{\text{input power} - \text{losses}}{\text{input power}} \qquad [3.7]$$

or alternatively:

$$\text{Efficiency } (\eta) = \frac{\text{output power}}{\text{output power} + \text{losses}} \qquad [3.8]$$

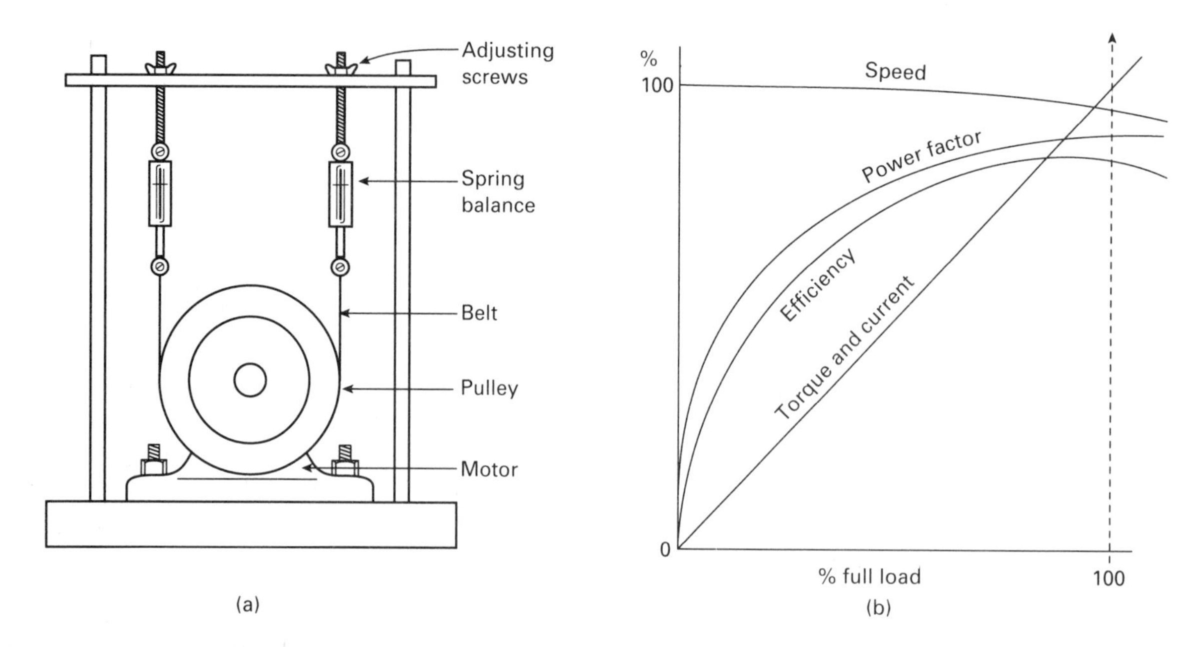

Figure 3.3 Induction motor load test (a) brake apparatus, (b) test results

Example 3.1

Figure 3.3 shows the results of a load test on a small single-phase, cage induction motor using a spring balance brake. From the data given below determine the motor's efficiency:

Supply voltage	110V
Load current	3 A
Power factor	0.96 lagging
Speed	23.8 rev/s
Radius of pulley	0.08 m
Spring balance difference	18 N

Solution

Electrical input power

$$P_I = VI \times \text{p.f.}$$
$$= 110 \times 3 \times 0.96$$
$$= 316.8 \text{ W}$$

Mechanical output power

$$P_O = 2\pi nT$$
$$= 6.28 \times 23.8 \times (18 \times 0.08)$$
$$= 215.22 \text{ W}$$

Therefore

$$\% \text{ efficiency } \eta = \frac{\text{output power} \times 100}{\text{input power}}$$
$$= \frac{P_O \times 100}{P_I}$$
$$= \frac{215.22 \times 100}{316.8}$$
$$= 68\%$$

Example 3.2

A 240 V single-phase a.c. induction motor with a power factor of 0.7 lagging, drives a lathe fitted with a cutting tool which exerts a force of 200 N on a work piece 10 cm in diameter. If the motor runs at a speed of 24.67 rev/s and has an efficiency of 82%, what is its electrical input power and supply current?

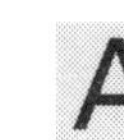

Solution

$$\text{Torque} \quad T = Fr$$
$$= 200 \times 0.05$$
$$= 10 \text{ Nm}$$

Output power

$$P_O = 2\pi nT$$
$$= 2\pi \times 24.67 \times 10$$
$$= 1550 \text{ W}$$

Efficiency $\eta = P_O/P_I$

Input power

$$P_I = P_O/\eta$$
$$= 1550/0.82$$
$$= 1890 \text{ W}$$

but $\quad P_I = V \times I \times p.f.$

therefore $I = P_I/(V \times p.f.)$

$$= 1890/(240 \times 0.7)$$
$$= 11.25 \text{ A}$$

Example 3.3

a) A motor-generator set runs at speed of 20 rev/s and delivers a current of 27 A to an external load. If the motor's electrical input is 2 kW and its copper and magnetic losses are 120 W and 80 W respectively, what is its efficiency, ignoring other losses?
b) What is the motor's torque?
c) What is the generator's voltage if its efficiency is 75%?

Solution

a) The motor's mechanical output power is its electrical input power minus the losses:

$$P_O = P_I - P_C - P_M$$
$$= 2000 - 120 - 80$$
$$= 1800 \text{ W}$$

b) Since

$$P_O = 2\pi nT$$

then $$T = \frac{P_O}{2\pi n}$$
$$= \frac{2000}{2\pi \times 20}$$
$$= 15.9 \text{ Nm}$$

c) The mechanical output of the motor is the generator's input.

Since $\eta = \dfrac{P_O}{P_I}$

then $\quad P_O = \eta \times P_I$

$$= 0.75 \times 1800$$
$$= 1350 \text{ W}$$

The generator's terminal voltage is:

$$V = P/I$$
$$= 1350/27 = 50 \text{ V}$$

Example 3.4

A 415 V, 50 Hz three-phase cage induction motor has an output rating of 12 kW. Determine its efficiency if its full-load current is 33 A and power factor is 0.7 lagging.

Solution

$$\% \text{ Efficiency } \eta = \frac{\text{output power rating}}{\text{electrical input power}}$$
$$= \frac{12{,}000}{\sqrt{3} \times 415 \times 0.7 \times 33}$$
$$= 72\%$$

Example 3.5

A 6-pole, 415 V, 50 Hz three-phase induction motor has a power factor of 0.7 lagging and is used to drive an elevator lifting 100 kg at a rate of 3.6 m/s. If the elevator and motor efficiencies are 75% and 85% respectively, determine for full load conditions the motor's output power and input power.

Solution

The power required by the elevator is given by the formula:

$$\text{Power} = \frac{\text{Work done}}{\text{time}}$$

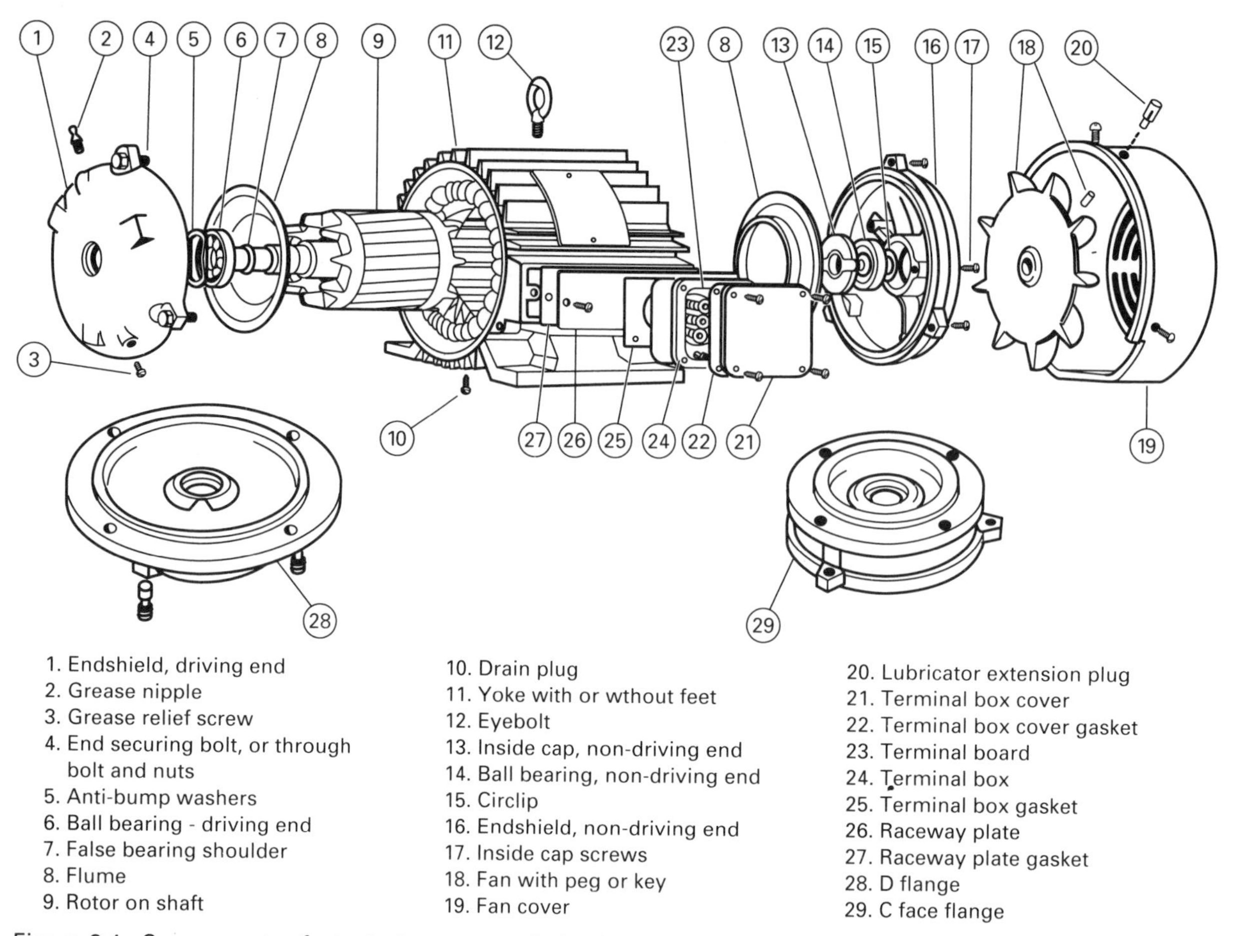

Figure 3.4 Components of a typical cage-rotor induction motor

$$= \frac{100 \times 9.81 \times 3.6}{1}$$

$$= 3531.6 \text{ W}$$

The motor's output is the elevator's input since both machines are coupled together on the same shaft. Thus:

Motor's output power

$$= \frac{\text{Elevator's input power}}{\text{Efficiency}}$$

$$= \frac{3531.6}{0.75}$$

$$= 4708.8 \text{ W}$$

Motor's input power

$$= \frac{\text{Motor's output power}}{\text{Efficiency}}$$

$$= \frac{4708.8}{0.85}$$

$$= 5539.76 \text{ W}$$

Cage induction motor

The construction of this type of motor is shown in Figure 3.4. Its stationary part the yoke or frame is a hollow cylinder made from welded or rolled steel supporting a slotted laminated stator core comprising annular silicon-steel stampings. On the inner circumference of the stator core in the slots is placed a distributed stator winding, insulated according to the supply voltage. The windings are either a single layer (one coil side per slot) or double layer (two coils per slot) arrangement with the ends brought out to a terminal box on the motor's casing. If the motor is designed for three-phase operation the three windings ends will be marked U1–U2, V1–V2 and W1–W2. Some motors are internally star connected and only have one end per phase brought out. For operation on a single-phase supply the motor's main winding will be marked U1–U2. If it has an auxiliary winding to facilitate starting it will be marked Z1–Z2. Other forms of lettering may be used, depending on the type of motor and its method of starting.

The motor's rotating part the rotor, consists of a laminated steel core having longitudinal slots into which are inserted lightly insulated copper or aluminium conductors called rotor bars. These rotor bars are short-circuited at each end by heavy copper or aluminium end rings which form a closed circuit reassembling a cage (hence the term cage rotor induction motor). The rotor shaft is supported at both ends by bearings and endshields and it may have attachments such as an internal fan to keep the windings cool or centrifugal switch which operates an auxiliary stator winding.

In a cage induction motor there is no electrical connection between the rotor and stator. The clearance (air-gap) separating both these parts is very small (about 1.5 mm for small motors and 2.5 mm for large motors).

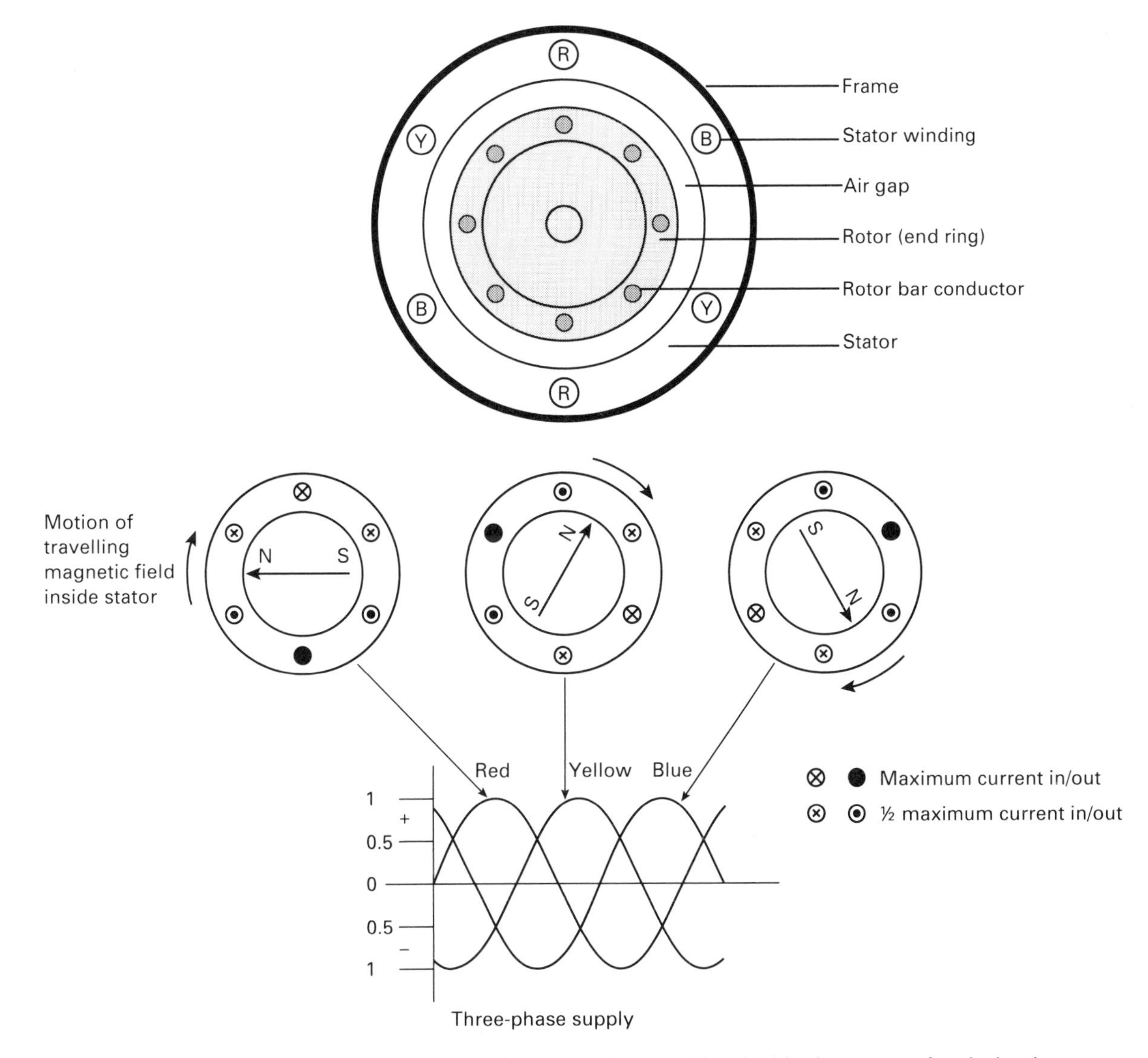

Figure 3.5 Rotating field produced by a three-phase supply, travelling inside the stator of an induction motor at synchronous speed

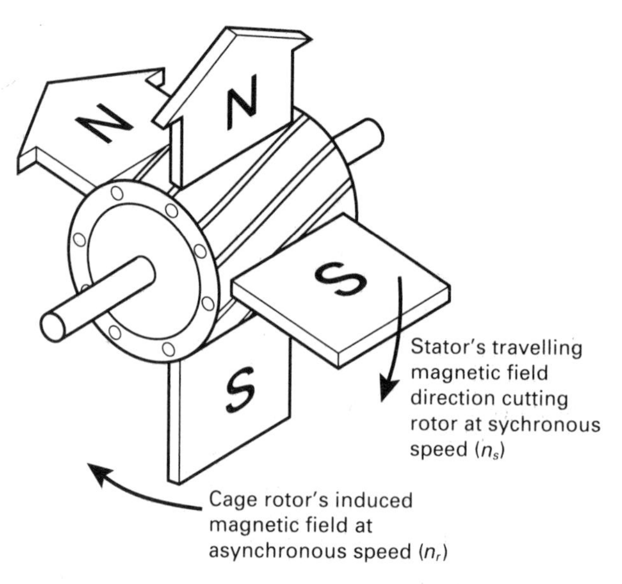

Figure 3.6 Field theory of induction motor based on the fact that like poles repel

Rotating magnetic field

It was explained in chapter two that a three-phase a.c. supply automatically produces a red, yellow and blue phase sequence with the phases displaced from each other by 120° (see Figure 2.30). This supply is ideal for operating a three-phase cage induction motor since its three stator windings will automatically produce a travelling magnetic field inside the stator core (see Figure 3.5).

The way in which this rotating magnetic field is created can be understood by considering the case of a motor having one slot per pole per phase. When current in the red phase winding is positive and at a maximum value, the yellow and blue phase windings are negative and half their maximum value. At this instance the resulting magnetomotive force (i.e. the ampere-turns required to produce magnetic flux) acts at right angles to the red phase winding. This creates a magnetic field that cuts across the cage rotor bars (see Figure 3.5). When the yellow phase winding reaches a positive maximum value the stator field moves 120°. Similarly, when the blue phase winding reaches a positive maximum value the field rotates a further 120°. This rotation of the magnetic field is responsible for inducing currents into the rotor bar conductors. The field travels at **synchronous speed** (n_s) and is shown in a clockwise direction. It is constant and governed by the frequency of supply (f) and the number of pairs of poles (p) in the motor's stator windings. It is expressed as:

$$n_s = f/p \text{ rev/s} \qquad [3.9]$$

With a supply frequency of 50 Hz and the stator only having one pair of poles per phase the magnetic field would travel at a synchronous speed of 50 rev/sec. For the same frequency a 4-pole machine would have a synchronous speed of 25 rev/sec.

Rotor operation

You will see from Figure 3.6 that the motor's rotor travels in the same direction as the rotating magnetic field. The rotor will receive full induced voltage at standstill and because its rotor bars are short circuited at both ends, induced currents flow along them. This gives rise to a rotor field system which interacts with the stator field system and the rotor begins to accelerate under the action of a net torque. As the rotor speed increases to nearly the same value as the synchronous speed, the rate of flux cutting becomes reduced and the rotor's induced voltage and current also reduce. This affects the net torque and any further increase of speed is not possible.

It should not be too difficult to see why the rotor cannot run as fast as the stator's magnetic field since no electromagnetic induction could occur. A type of motor having a rotor capable of running at synchronous speed is called a **synchronous motor**. It has a permanent magnet rotor which locks itself into the stator's rotating field. However, in the asynchronous motor, the difference between the rotor speed (n_r) and the stator synchronous speed (n_s) is called **slip** (s). It is proportional to the load on the motor, expressed in per unit as:

$$s = \frac{n_s - n_r}{n_s} \qquad [3.10]$$

When the rotor is not moving its slip is unity (i.e. $s = 1$). If synchronous speed was possible, its slip would be zero (i.e. $s = 0$). The motor normally runs at constant speed, operating with a slip between 0.03 and 0.05 per unit.

Example 3.6

A 4-pole cage induction motor operates from a 50 Hz supply and runs at 24.25 rev/s. What is its slip speed and per unit slip?

Solution

The motor's synchronous speed is given by formula [3.9], hence:

$$n_s = f/p$$
$$= 50/2$$
$$= 25 \text{ rev/s}$$

The slip speed is (25 – 24.25)

$$= 0.75 \text{ rev/s}$$

Per unit slip (s) is given by formula 3.10, hence:

$$s = \frac{n_s - n_r}{n_s}$$
$$= \frac{25 - 24.25}{25}$$
$$= 0.03 \text{ per unit } (3\%)$$

Example 3.7

The rotor of a 60 Hz cage induction motor runs at 19 rev/s with a 0.05 per unit slip. What is the motor's synchronous speed and how many poles are there on the stator?

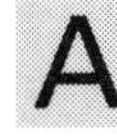

Solution

Transpose formula [3.10] and make n_s the subject of the formula.

Since $$s = \frac{n_s - n_r}{n_s}$$

then $$sn_s = n_s - n_r$$

and $$n_r = n_s - sn_s$$
$$= n_s(1 - s)$$

therefore

$$n_s = \frac{n_r}{1 - s}$$
$$= \frac{19}{0.95}$$
$$= 20 \text{ rev/s}$$

The number of poles is found from formula [3.9], making p the subject of the formula, hence:

$$p = f/n_s$$
$$= 60/20$$
$$= 3$$

Since p is pairs of poles, the stator has six poles.

Example 3.8

Calculate the rotor speeds of a three-phase, 2-pole cage induction motor for the following supply frequencies 20 Hz, 50 Hz and 120 Hz. Assume a fractional slip of 0.05 in each case.

Solution

Synchronous speed on 20 Hz:

$$n_s = f/p$$
$$= 20/1 = 20 \text{ rev/s}$$

Rotor speed at 5% slip:

$$n_r = n_s - \% \text{ slip}$$
$$= 20 - \frac{5 \times 20}{100}$$
$$= 19 \text{ rev/s}$$

Synchronous speed on 50 Hz:

$$n_s = 50/1$$
$$= 50 \text{ rev/s}$$

Rotor speed at 5% slip:

$$n_r = n_s - \% \text{ slip}$$
$$= 50 - \frac{5 \times 50}{100}$$
$$= 47.5 \text{ rev/s}$$

Synchronous speed on 120 Hz:

$$n_s = 120/1$$
$$= 120 \text{ rev/s}$$

Rotor speed at 5% slip:

$$n_r = n_s - \% \text{ slip}$$
$$= 120 - \frac{5 \times 120}{100}$$
$$= 114 \text{ rev/s}$$

When the cage induction motor is mechanically loaded its speed begins to fall and its slip increases. Its frequency and rotor current increase and, as a result, more current is taken by the stator windings. Figure 3.7 shows the torque-speed characteristic of a typical motor. The shaded area represents the normal operating region and the pull-out torque is approximately 2.0 to 2.5 times normal full-load torque. The motor's starting current is about 6–8 times full-load current. Both starting torque and starting current are related to rotor resistance. The higher the resistance the greater will be the starting torque since the starting current is low. The dotted line in Figure 3.7 represents an improved rotor cage to increase the starting torque.

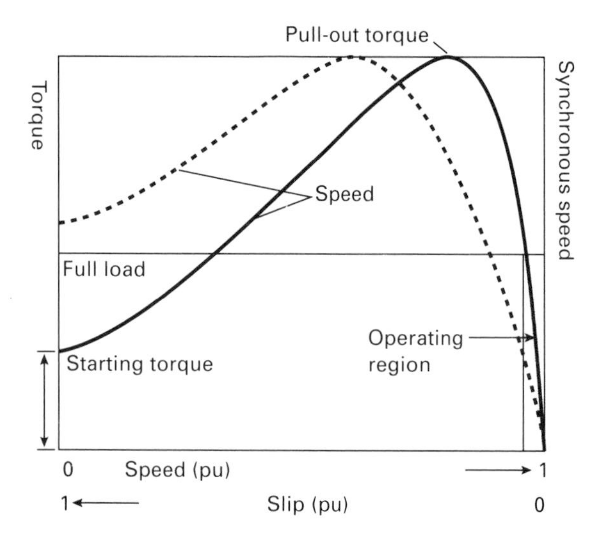

Figure 3.7 Torque-speed characteristics of two cage induction motors

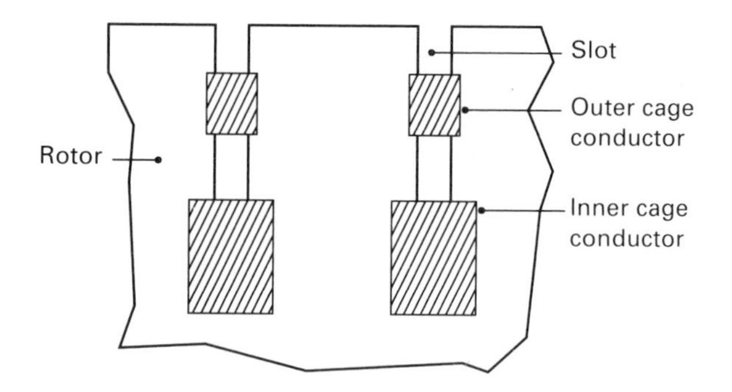

Figure 3.8 Double cage rotor

Various rotor bar shapes have been designed to improve the starting torque and one satisfactory method is by using a double cage rotor. This takes advantage of slip frequency decreasing as the motor speed increases. The rotor has an outer cage and inner cage. The outer cage is in shallow slots and the conductors are made of bronze having a relatively high resistance and low reactance. The inner cage is in deep slots and the conductors are made of high conductivity copper having relatively high reactance and low resistance (see Figure 3.8). The motor starts up on the high resistance cage and during acceleration its torque-speed characteristic changes over to the high reactance cage. The total torque being the sum of the two cages. This arrangement not only improves the starting torque but also improves the motor's performance at full-load speed.

Wound-rotor induction motor

Another method of improving starting torque is to provide external resistance to the motor's rotor.

This type of induction motor operates from a three-phase supply and is often called a **slip-ring motor**. Unlike the cage rotor type with its rotor bars short-circuited by two end rings, this motor has a wound rotor. One end of the winding is star connected while the other end is connected to three slip rings. You will see from Figures 3.9 and 3.10 that external rotor resistance is connected to the slip-rings via brushgear.

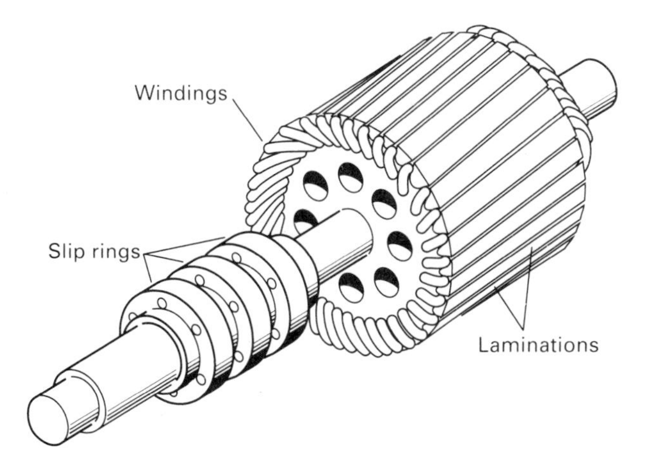

Figure 3.9 Three-phase wound-rotor induction motor

This resistance is progressively cut out as the motor speeds up until finally the rotor winding is short-circuited. It then becomes another star connection with both ends simulating the short-circuiting end rings of a cage rotor (see circuit diagram Figure 3.10).

You will see from this motor's speed-torque characteristic that a series of curves appear. Curve A illustrates maximum torque with all resistance in circuit while curves B and C represent intermediate steps of cutting resistance out. Curve D is the motor's normal operating characteristic. You will observe that it cuts the full load torque line (shown dotted) at point X which indicates 5% rotor slip.

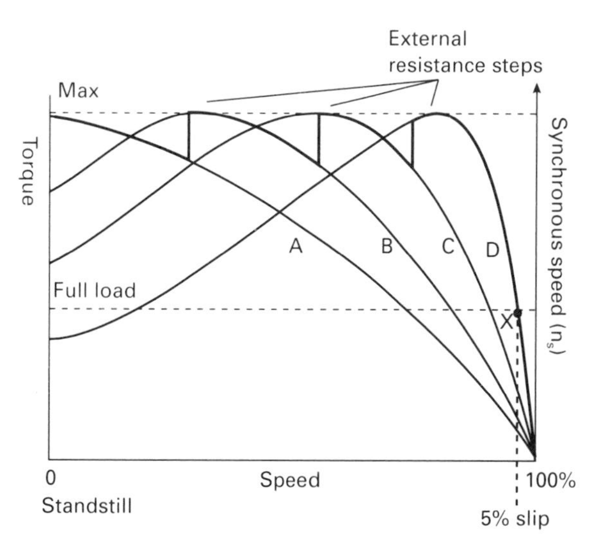

Figure 3.11 Three-phase wound-rotor induction motor speed/torque charateristics

In practice, a contactor-type starter is used to provide the necessary circuit protection such as no-volt protection and overload protection. The starter will also have an interlock facility to ensure that its contactor cannot be closed while the motor is at standstill until all rotor resistance is in circuit. To provide smoother starting, large slip-ring motors use liquid resistance starters containing caustic soda or washing soda.

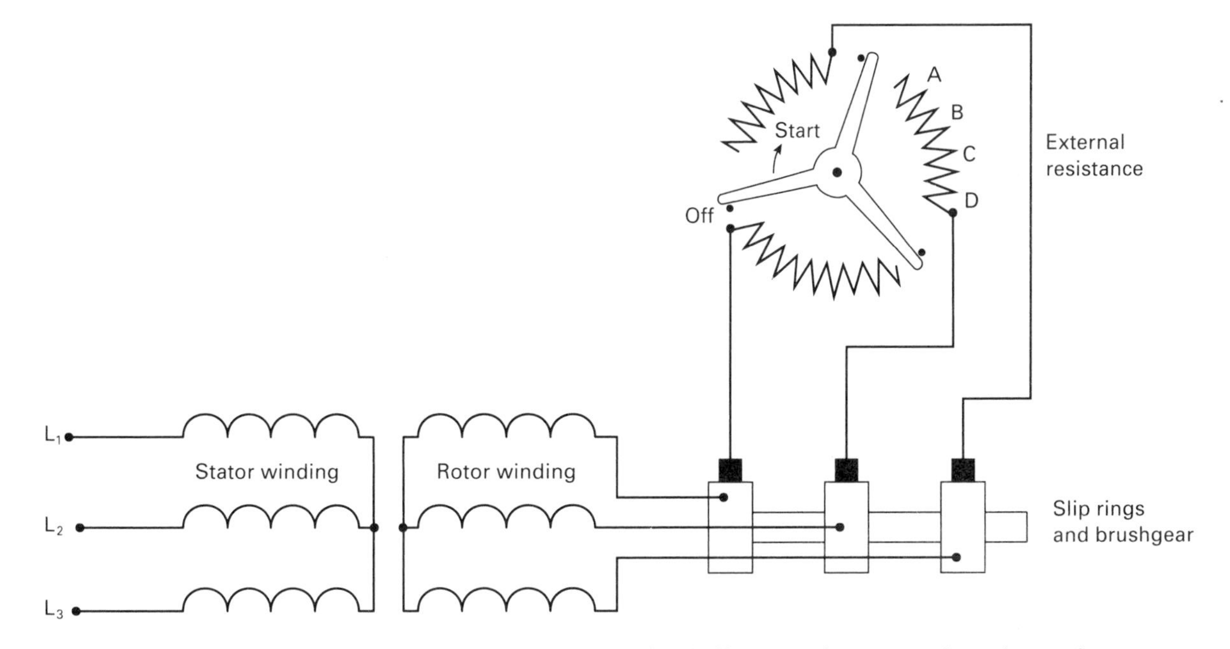

Figure 3.10 Three-phase wound-rotor induction motor; circuit diagram of motor and starting resistance to improve torque

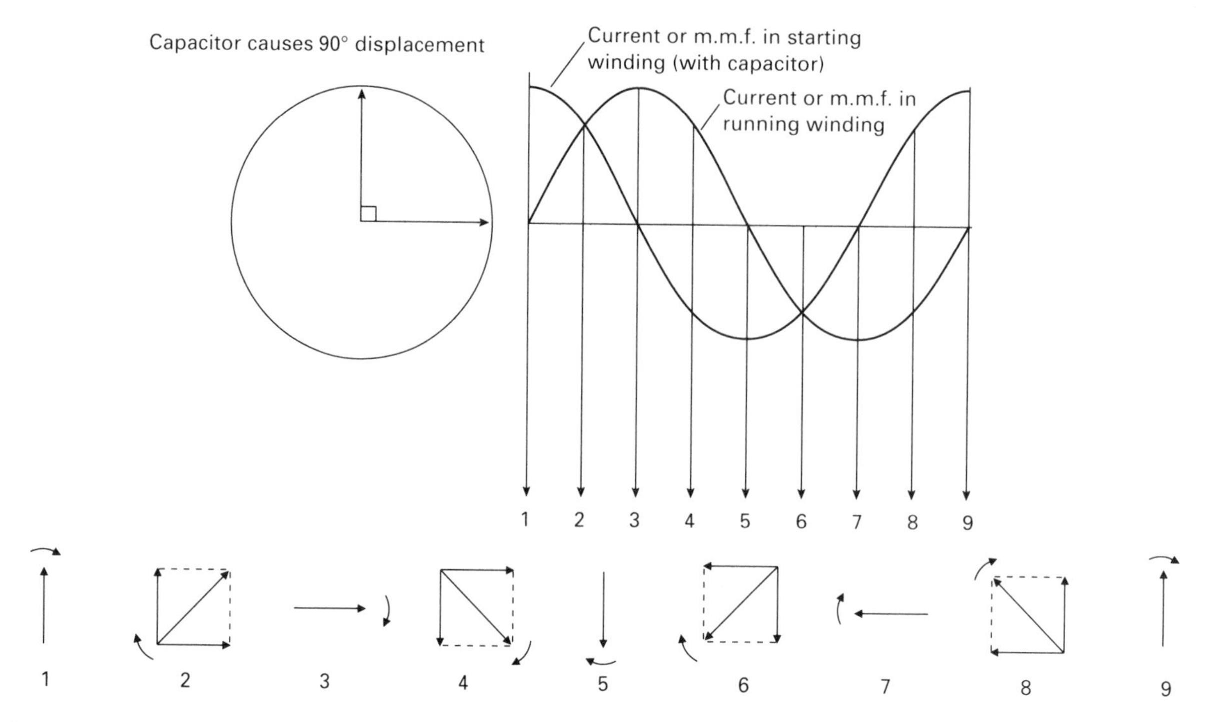

Figure 3.12 Split-phase rotation of stator fields

Application

With three-phase supplies available and because of their robust construction and constant speed characteristics, three-phase induction motors have wide acceptance in industrial and commercial processes. The majority are 4-pole machines running at 25 rev/sec (1500 rev/min) and are used for compressors, pumps and fans, etc. Slip-ring motors are normally chosen for output ratings above 100 kW. Reversal of direction is by interchanging any two of the stator winding connections which actually reverses the stator's travelling magnetic field direction.

Single-phase induction motors

The basic construction of these motors is similar to that of three-phase machines. One might assume that a single-phase motor only requires one stator winding for operation but unfortunately this is not the case. A single-phase supply connected to one stator winding would not produce a rotating magnetic field; it would only set up a pulsating field similar to the primary winding of a single phase transformer. The rotor would be unable to produce a starting torque.

One method of overcoming the problem is in the design of a **split-phase motor** which has two separate stator windings. The windings are initially connected in parallel and arranged spatially to be 90° apart. One of the windings is called a **running winding** and the other is called a **starting winding**. The running winding is the main winding and is always connected to the supply but the starting winding (sometimes called auxiliary winding) is designed to be disconnected from the circuit.

To enable the stator to simulate a rotating magnetic field, one of the windings has a different resistance (R) value or reactance (X) value. By doing this both windings carry current which are out of phase with each other as was explained in chapter two. This is the meaning of split-phase. The resultant magnetic field produced by these currents will create the necessary torque on the rotor to make the motor self-starting. Figure 3.12 shows the production of the rotating field. Two motors operating on this principle are called resistance-start and capacitor-start.

The **resistance-start motor** (see Figure 3.13(a)) has more resistance in its starting winding than in its running winding. It is usually wound with the same number of turns as the main winding but the wire is of a smaller cross sectional area. An alternative arrangement is to connect a resistor in

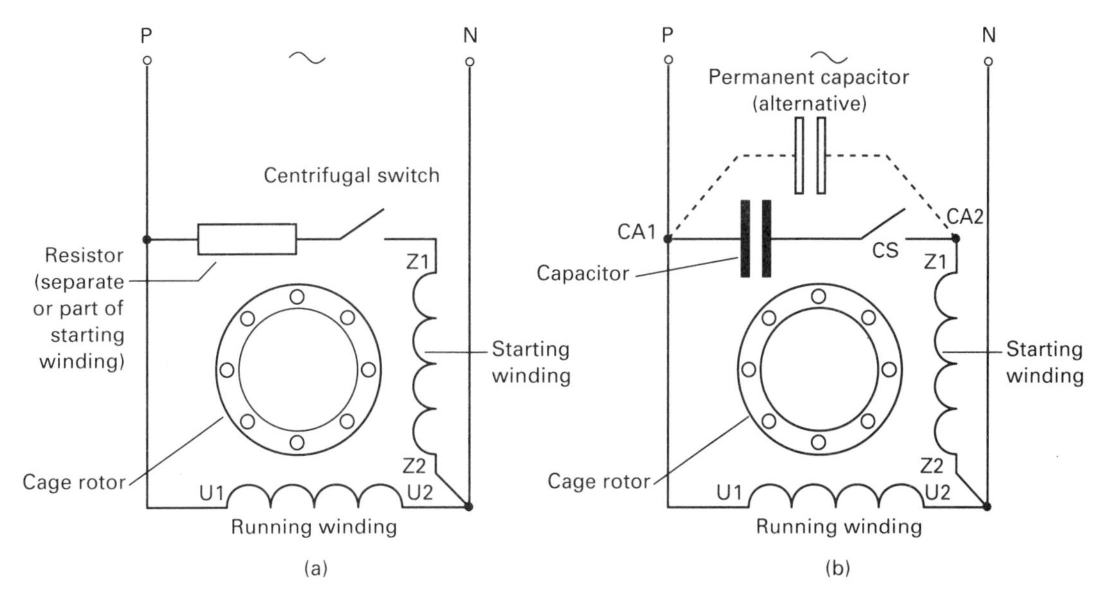

Figure 3.13 Types of split-phase induction motor (a) resistance start, (b) capacitor start

series with the starting winding. The current in the starting winding lags behind the supply voltage by approximately 30° while current in the main winding lags behind the supply voltage by approximately 70°.

To reduce copper losses the starting winding is disconnected from circuit when the motor reaches approximately 75–80% of its full load speed. The device which achieves this is called a **centrifugal switch** and is attached to the motor's shaft as shown in Figure 3.14. You will see from the circuit diagram that the switch is wired in series with the starting winding. For motor ratings up to 250 W, starting torque is approximately 1½ to 2 times full-load torque. Power factor and efficiency are approximately 0.7 and 55% respectively.

The **capacitor-start motor** (see Figure 3.13(b)) uses a paper dielectric capacitor for continuous operation or a dry-type electrolytic capacitor for non-continuous operation. Size of capacitor

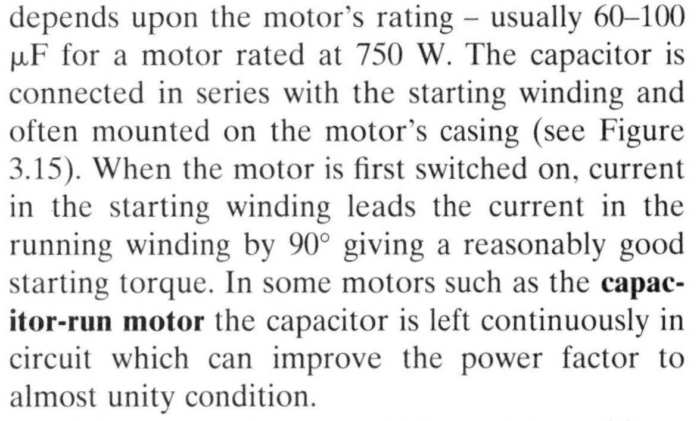
depends upon the motor's rating – usually 60–100 μF for a motor rated at 750 W. The capacitor is connected in series with the starting winding and often mounted on the motor's casing (see Figure 3.15). When the motor is first switched on, current in the starting winding leads the current in the running winding by 90° giving a reasonably good starting torque. In some motors such as the **capacitor-run motor** the capacitor is left continuously in circuit which can improve the power factor to almost unity condition.

Split-phase motors are widely used for refrigeration equipment, fans, pumps and small power application, where a constant drive is required. Reversal of direction is achieved by changing over

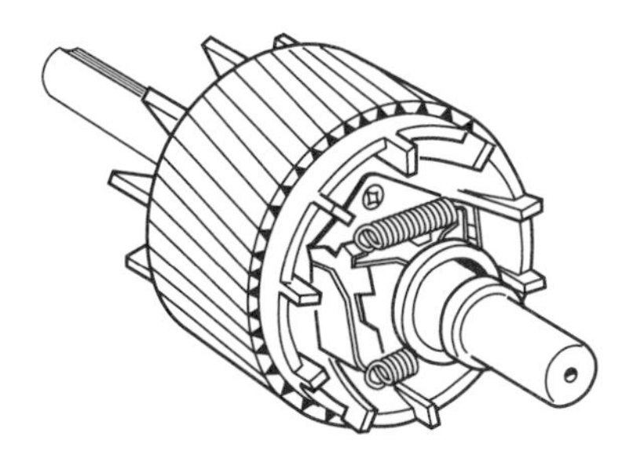

Figure 3.14 Cage rotor showing centrifugal switch mechanism

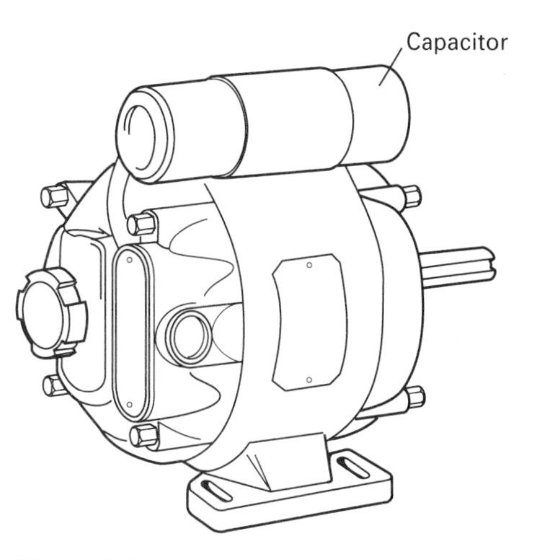

Figure 3.15 Capacitor motor

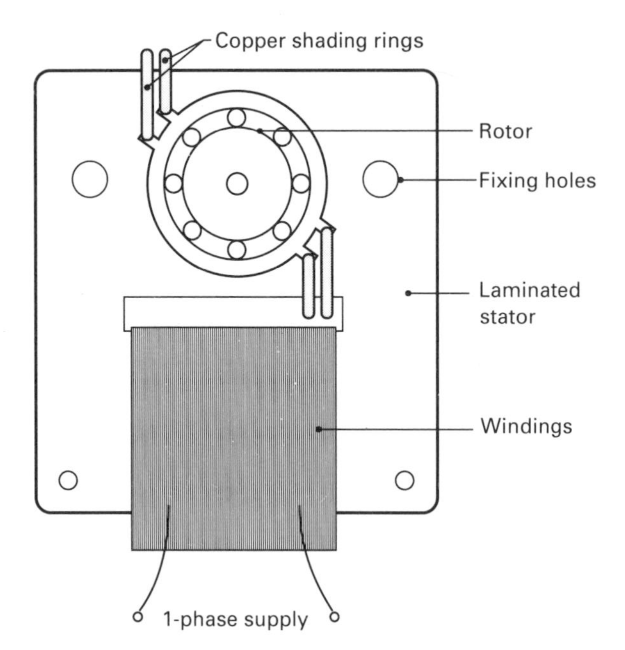

Figure 3.16 Shaded-pole (unicoil) motor

the starting winding connections. Another type of cage induction motor with a stator modification to operate from a single-phase supply is called the **shaded pole motor**.

Figure 3.16 is a **unicoil motor** having a wound coil on one side of its laminated stator core and two copper shading rings on the other side. The shading rings create an artificial phase shift, sufficient to produce a weak starting torque for driving oven fans, timers, etc, up to about 50 W.

Another type with a higher rating is called a **salient-pole motor**. This has prominent stator poles which are again fitted with copper shading rings (see Figure 3.17). The shading rings act as short-circuited, low-resistance coils. When the supply is switched on, eddy currents are induced in the shading rings and they produce their own magnetic field. The flux produced by this field opposes the main flux and you will see that it concentrates on one side of the pole (see Figure 3.18). This action causes a slight delay before the flux in the shading area reaches a maximum value. When the main flux starts to decrease, the field inside the shading ring increases, giving the effect of sweeping from one side of the pole to the other. This shift is sufficient to create a torque to make the motor self-starting. Motors of this type are often 4-pole machines having output ratings around 125 W which are used for constant speed applications.

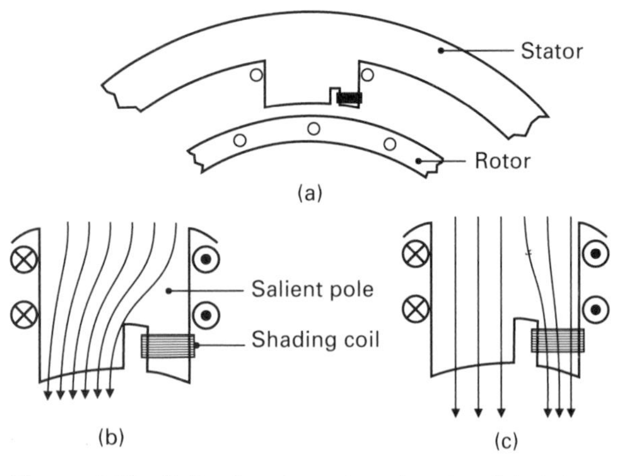

Figure 3.18 Salient-pole motor showing (a) stator and rotor, (b) shading pole opposing main field, (c) shading pole aiding main field

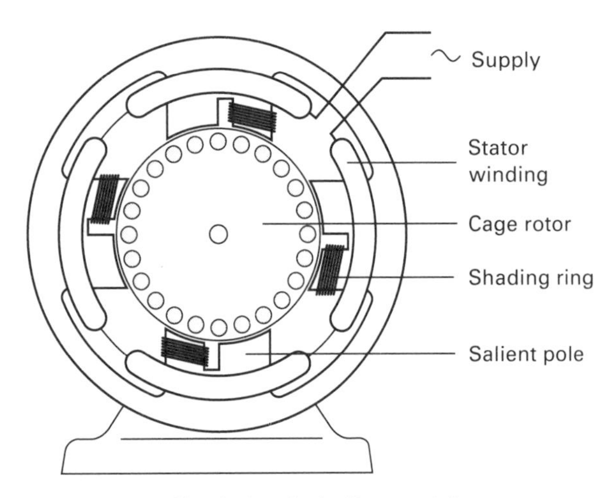

Figure 3.17 Shaded-pole (salient-pole) motor

Universal motor

This motor is constructed on the lines of a d.c. series motor having a fixed field winding on a

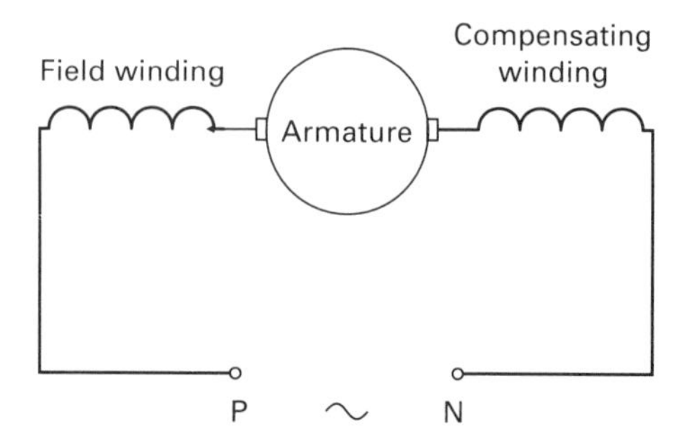

Figure 3.19 Circuit diagram of a universal motor

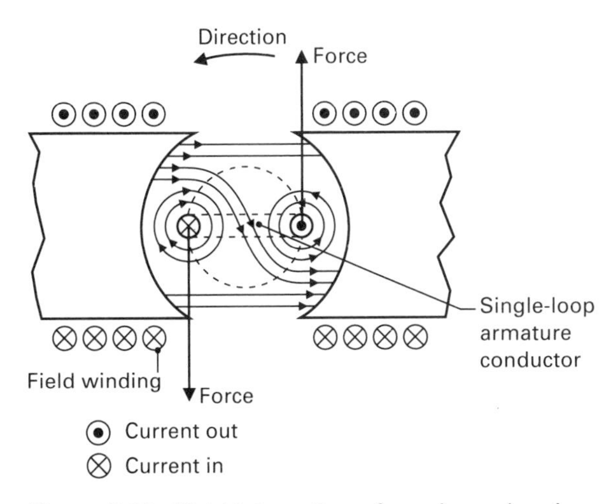

Figure 3.20 Field interaction of a universal series motor

laminated yoke. The field winding is connected in series with an armature through a commutator and brushgear. The armature has windings.

Figure 3.19 shows a wiring diagram of the motor and Figure 3.20 shows the field interaction between the main field and armature field to produce rotation of the armature. In this diagram the poles of the electromagnetic are shown projected. For a.c. operation a bridge rectifier could be inserted to allow the motor to run as a d.c. series motor.

Alternatively, the main field winding is distributed around the stator in slots similar to that of the induction motor and the motor connected directly to the a.c. supply. It should be mentioned that the a.c. supply sets up eddy currents, causing I^2R losses in the core. For this reason all ferromagnetic parts need to be highly laminated. The motor suffers from sparking at the brushes which rest on the armature's commutator. It is a problem created by transformer action from the stator field and causes induced e.m.f.'s to track across the brushes and commutator. Motors designed with high kilowatt ratings are often fitted with a series compensating winding to counteract this problem.

The universal motor's operating characteristics are shown in Figure 3.21. It will be seen that on light loads the speed is high. For this reason it should always be connected to some form of load to avoid racing up to a dangerous speed. Since the supply is alternating current, both field polarity and armature polarity change together in the same time instants and the motor will run in the same direction. Reversal of rotation is achieved by either changing over the main series field connections or changing over the armature field connections. The universal motor has considerable application in domestic appliances such as washing machines, vacuum cleaners, hair dryers, hand mixers, etc.

Motor starters

The prime function of a motor starter is to connect the motor to the supply and accelerate it to full speed without disturbance in the line or to other connected machinery or equipment. The starter may also provide the following functions:

1) to disconnect the motor from the supply in the event of an overload or fault within the motor;
2) to disconnect the motor from the supply for safety reasons – protecting the operator or other person(s) – achieved by incorporating various safety devices which trip out the starter;
3) to control the stopping and starting operations of the motor to meet various requirements of the work process.

The problem associated with the starting and acceleration of a motor, particularly three-phase induction motors, arises from its speed-torque characteristic. It was explained earlier that both starting torque and starting current are related to the motor's rotor resistance. While the former may be a fairly modest value (150% of the full load torque) the latter can be very high, approaching 6–8 times full load current. These two factors are

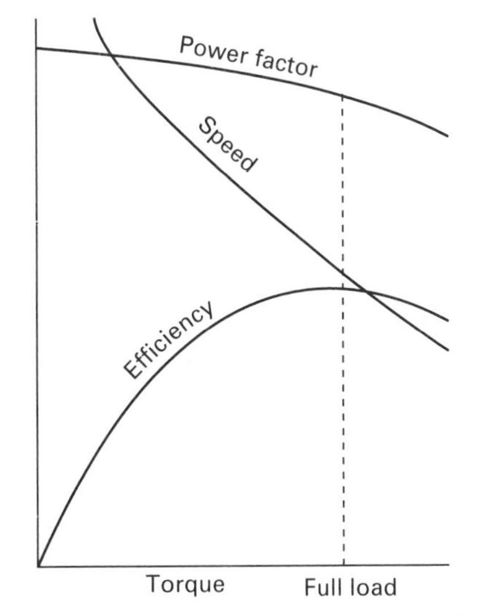

Figure 3.21 Performance curves for universal series motor

further related to the motor's terminal voltage. Various types of starter are designed to reduce the voltage to the motor during the initial starting period. Once the motor is running at its rated speed the voltage can be restored to its supply value. In the case of motors of fractional kilowatt rating and where no appreciable disturbance on the supply lines is likely to occur, or where consent is given by a supply authority, then direct-switching of the motor at full voltage is possible. Some common starting methods are given below with obvious emphasis on three-phase supplies.

Direct-on-line starter

This is the simplest and least expensive form of starter for small motors up to about 3.7 kW especially for motors connected to light loads. With heavier loads the running up speed is prolonged and the motor may suffer damage from overheating caused by the high starting current. Very small motors of rating not exceeding 0.37 kW, may be switched onto the supply without any need of overload protection (see BS 7671 and IEE Wiring Regulations, Reg. 552–01–02).

Where it is necessary for the control circuit to provide some of the functions mentioned earlier, such as stopping and starting, overcurrent and no-load protection then a direct-on-line (DOL) contactor-type starter is employed. Figure 3.22 shows a circuit diagram of this starter controlling a three-phase induction motor. The circuit shows supply isolation, circuit protection in the form of overloads in each line and no-volt protection by means of the contactor's operating coil. You will see that facilities exist to start and stop the motor remote from its working position. This requires start buttons to be wired in parallel and stop buttons to be wired in series with the no-volt coil.

A number of modifications can be made to the circuit such as operation on a single-phase supply to control a single-phase motor as in Figure 3.23 as well as inching, external interlock facility and alarm circuit.

Star-delta starter

Figure 3.24 shows the connections of a hand-operated **star-delta starter**. It requires an induction motor with six winding ends brought out of its

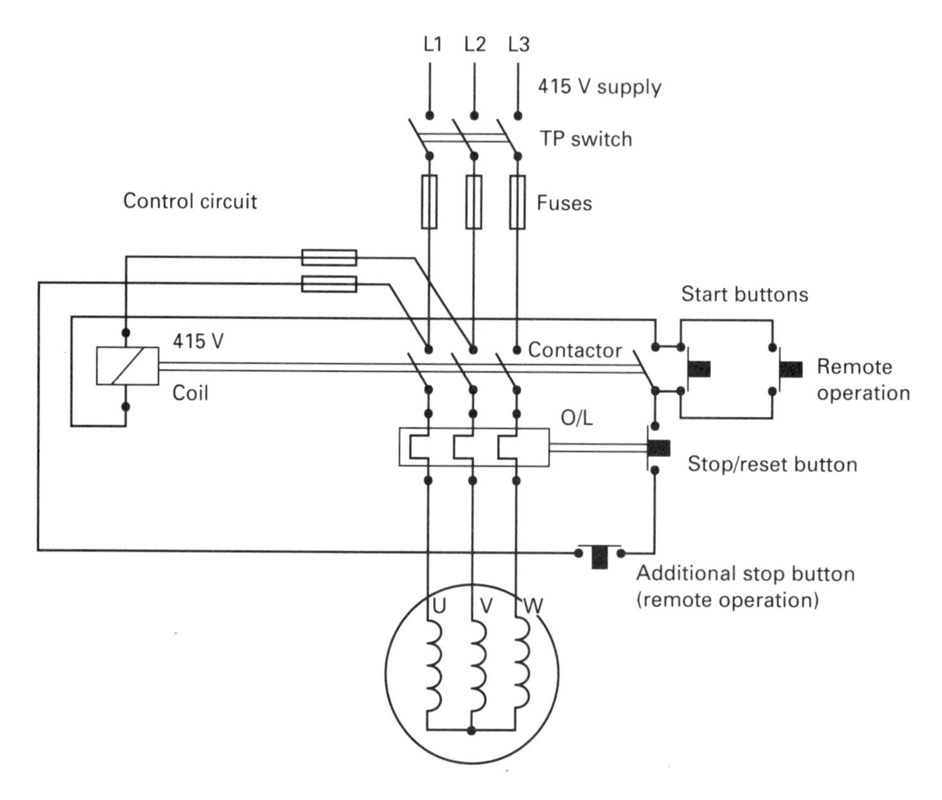

Figure 3.22 3-Phase motor with direct-on-line starter circuit

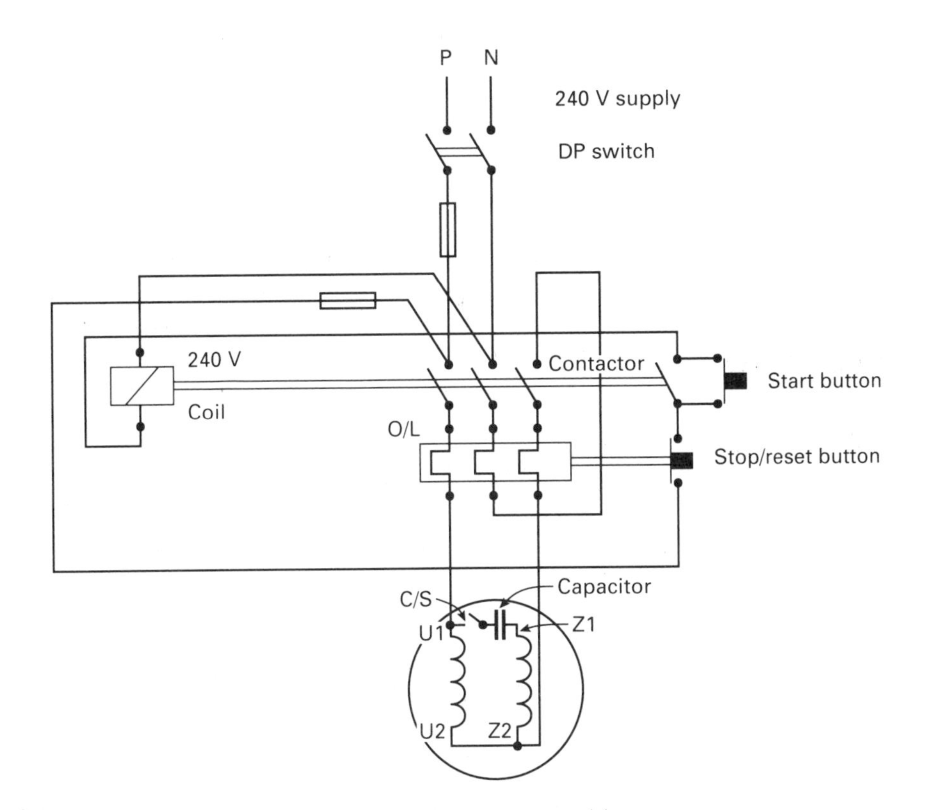

Figure 3.23 Circuit diagram of DOL starter for single-phase induction motor

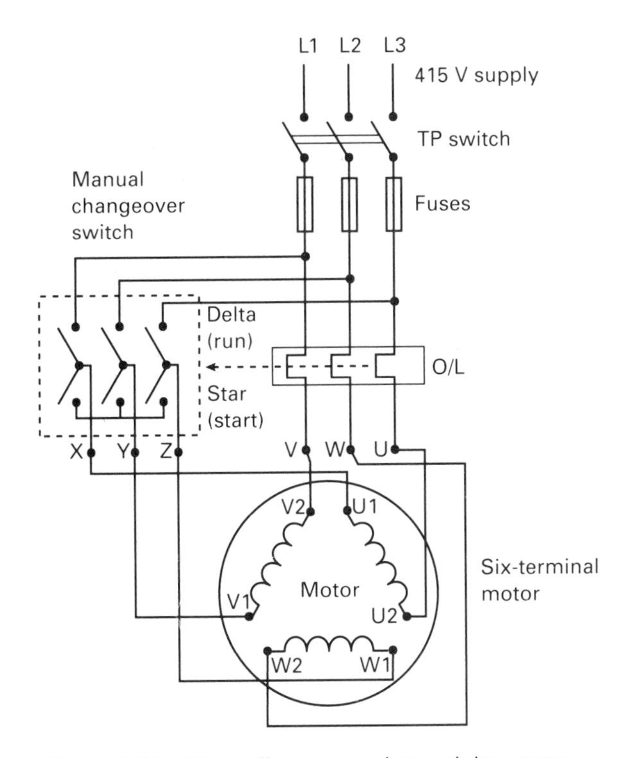

Figure 3.24 Manually-operated star-delta starter for induction motors

terminal box. You will see by tracing the circuit that in the start position of the changeover switch one end of the stator windings is connected in star. This allows the supply voltage to be applied across two windings in series and the motor will run up to speed at a reduced voltage. When the motor reaches its full speed, the manual changeover switch is moved to the run position and the motor operates in delta, each stator winding receiving full voltage.

The supply voltage per phase in star is only 58% of the value which would be applied if the windings were connected in delta (i.e. 240 V/415 V) and this reduces the starting torque and starting current to one-third. With only 50% starting torque it should be appreciated that this starting method is limited to light and medium loads such as centrifugal pumps and fans having low inertia.

Figure 3.25 shows a circuit diagram of an automatic star-delta starter. A schematic circuit has been drawn separate to the main circuit for clarity reasons (Figure 3.26) which involves a time delay that allows the star contactor S to operate before the delta contactor D. Briefly, when the start button is pressed, star contactor coil S is energised via

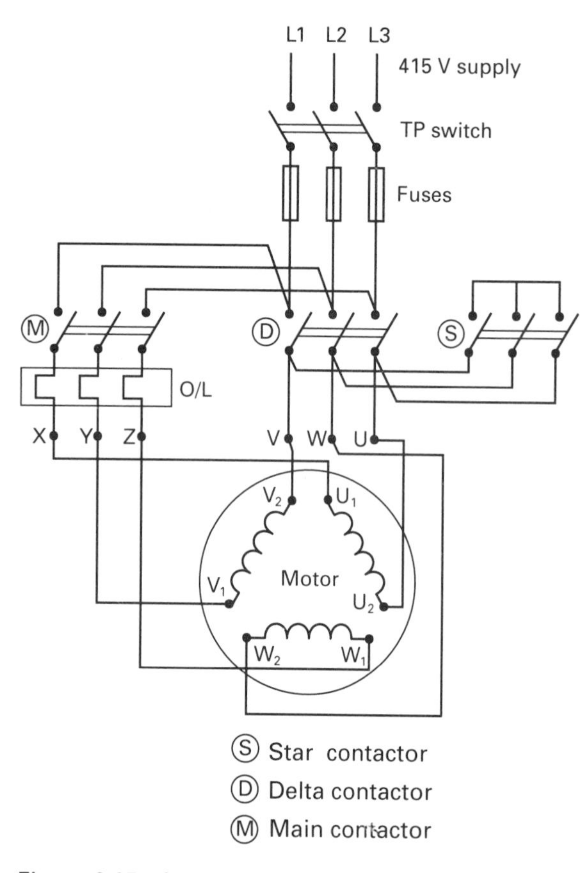

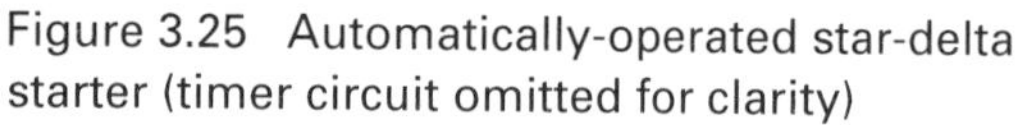

Figure 3.25 Automatically-operated star-delta starter (timer circuit omitted for clarity)

the normally closed contact D and timer contact T. This closes contacts S and M and the timer coil T operates. At the same time as coil S closes, current flows through the main contactor coil M. There is an interlock between the star and delta main contactors but after a short period of time the timer contacts open and current then flows through the normally closed star contact S to operate the delta coil. The motor windings will then receive full voltage.

Autotransformer starter

This can be a two- or three-stage voltage reduction method used where current limitation is required at starting. Unlike the star-delta starter it only requires three stator connections. In practice the autotransformer is star connected and tappings are taken off each phase winding to provide 40, 60 and 75% of the supply voltage. With starting torque being proportional to the square of the motor voltage this will give 24, 54 and 84% of the direct switching full-load torque.

These percentage torque calculations are based on the following:

Since $T \propto V^2$

When the transformer supplies 100% full voltage at 415 V, its full-load torque is 150%. For a tapping of 60% (249 V) its percentage torque will be:

$$T = \frac{150 \times 249^2}{415^2}$$

$$= 54\%$$

The initial starting current is 1.6, 3.6 and 5.6 times the full load current. Figure 3.27 shows a diagram of a manually-operated starter, although automatic starters are also available. The one illustrated shows the changeover switch in the start position and the transformer on its lowest tappings to allow the motor to gain speed, taking the least current. Moving the switch to the run position connects the motor to the full supply voltage.

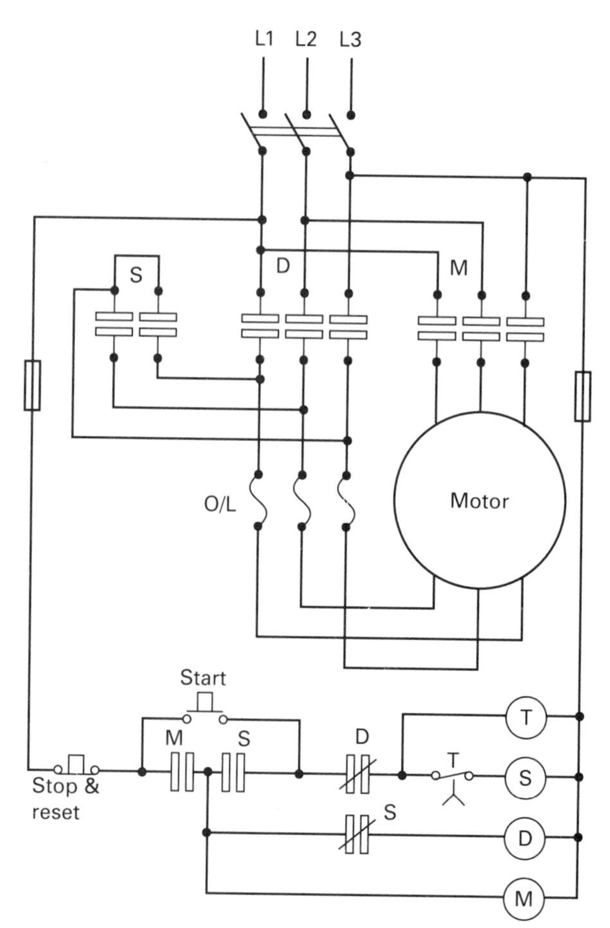

Figure 3.26 Control circuit with time delay for Figure 3.25

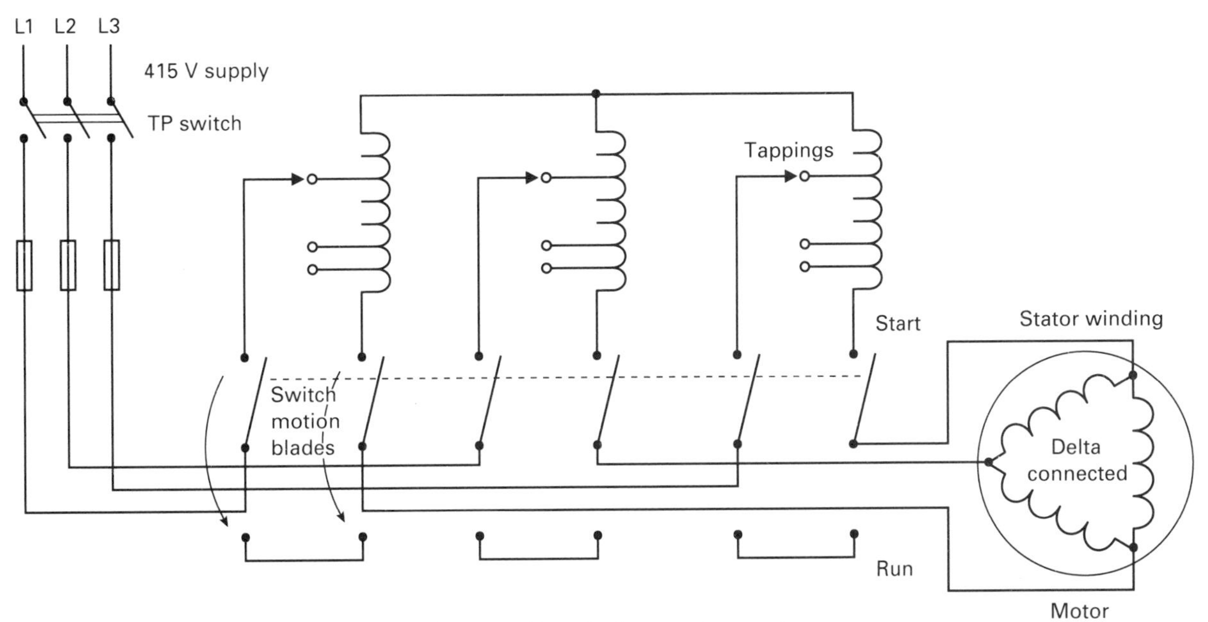

Figure 3.27 Manual changeover switch controlling 3-phase autotransformer

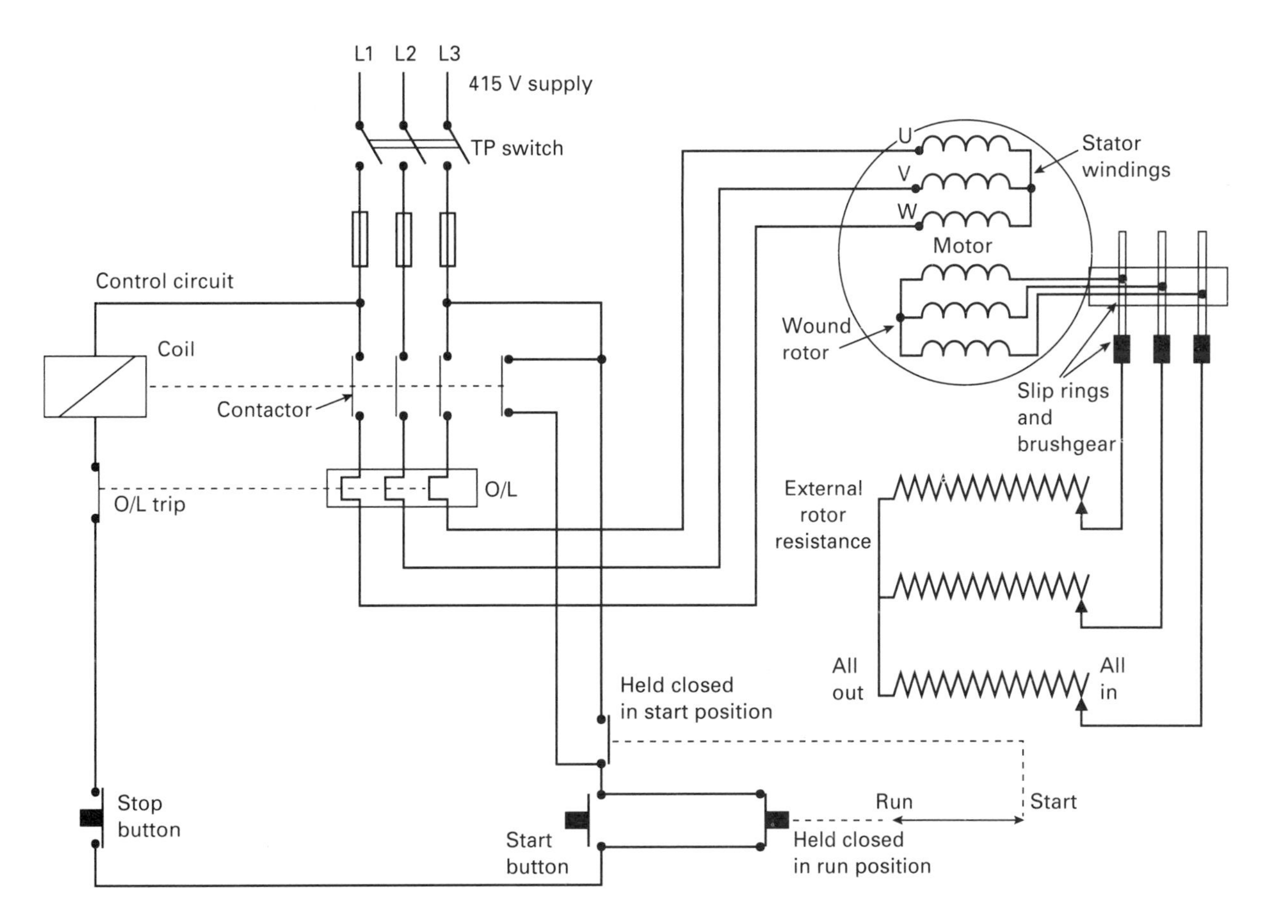

Figure 3.28 Contactor starter controlling three-phase, wound rotor, slip-ring motor

Rotor resistance starter

A diagram of this circuit is shown in Figure 3.28. You will see that circuit protection and control is by a separate contactor circuit and you will also notice that mechanical interlocks are inserted between the stator's contactor and the external rotor resistance. This ensures that the latter is all in circuit before the motor can be started. With maximum resistance in circuit, maximum torque is developed from standstill. As the speed of the slip-ring motor increases its torque drops off until it is balanced by the load and a steady speed is attained. Some resistance is then cut out and the motor's speed again increases until it levels out. This operation is repeated several times until all the external resistance has been removed and the motor finally runs at its rated speed. In this way the external rotor resistance is able to provide a certain amount of speed control but reduction in speed is accompanied by reduced efficiency. When running normally, the starter arm forms a short circuit across the slip-rings but with some motors a device at the slip-rings is used which allows the brushes to be lifted to minimise wear.

This method is ideal for motors up to 100 kW but in order to reduce any disturbance on the mains and to avoid vibration and shock to the driven machine the starting torque is often kept to approximately 150% of full load torque.

Motor protection

Protective devices are usually incorporated in a starter to protect the motor against overcurrent, loss of voltage, or as was seen above in the slip-ring motor, means to prevent misuse of the starter having interlock facilities. Overcurrent is often an overload caused by the motor taking too much current, either through stalling on load or by single-phasing (when one phase of a three-phase supply becomes open circuit). It leads to overheating of the windings and the eventual breakdown of insulation. To prevent this happening, use is made of an overload device which is often pre-set and operates by opening contacts which are wired in series with the starter's contactor coil. Obviously, an overload device set to trip slightly over the normal rating would be a nuisance and it is usual to find such devices set around 25% of the motor's rated current.

In practice, two types of overload device are

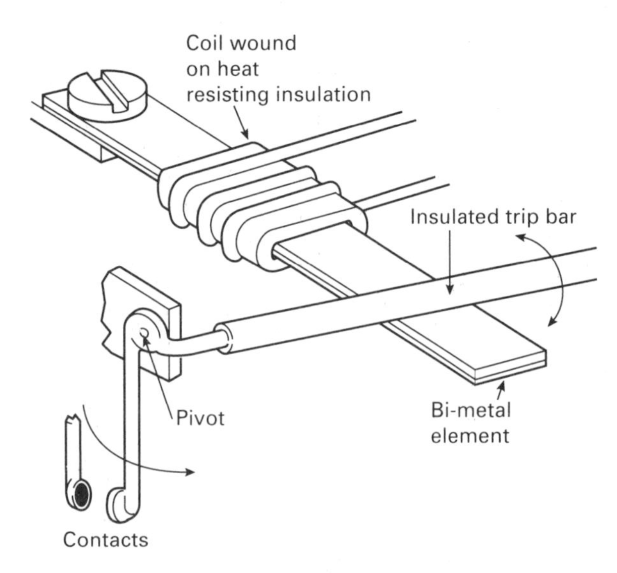

Figure 3.29 Bi-metallic overload device

used, namely, a thermal release or a magnetic release. The **thermal release** is used mainly in small contactor starters and consists of two metals having different coefficients of expansion (see Science Book 1 by the same author) and takes the form of a bimetal strip. When current is sufficiently strong enough in the motor circuit the heating effects

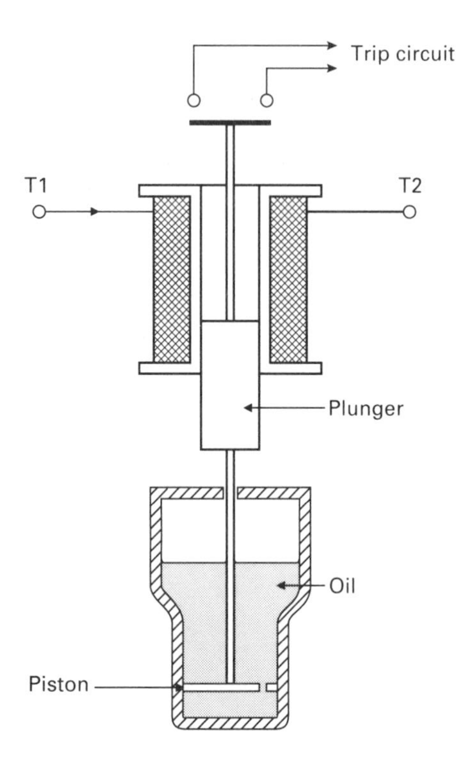

Figure 3.30 Oil-filled dashpot magnetic overload System

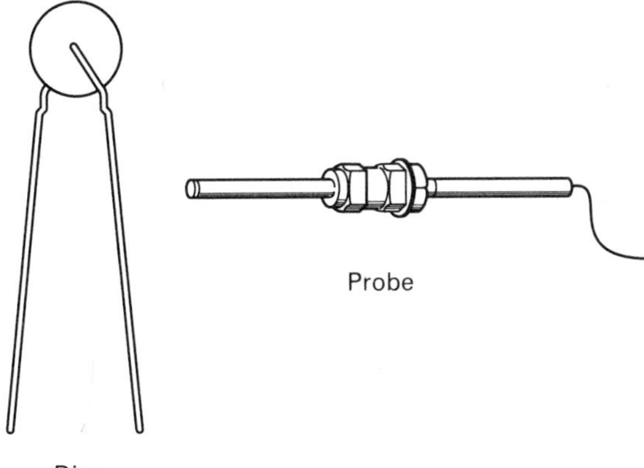

Figure 3.31 Thermistors

produced by it causes unequal expansion of the two strips and they bend. This action is designed to open contacts and de-energise the contactor coil of the control circuit. (See Figure 3.29)

The **magnetic release** is used with larger contactors and comprises a coil wired in the line circuit which automatically trips out the starter in response to an instantaneous overload. The device contains a plunger whose movement is damped by a dashpot which is oil-filled. This action avoids unnecessary tripping on short time overloads. Figure 3.30 shows a typical oil-dashpot.

Another method of protection for a motor is a device called a **thermistor** which is a thermally-sensitive semiconductor resistor (see Figure 3.31). The thermistor exhibits a positive temperature coefficient (ptc) with its resistance designed to change rapidly with changes in temperature. It is ideal for giving protection against excessive overheating of a motor's windings and is often connected in series to operate a relay or alarm circuit.

In the event of supply failure or the motor being switched off whether manually by the off button or automatically via overloads, the operating coil acts as a no-volt release to de-energise the contactor. This is an important safety feature for the operator of the motor and often the coil is wired through a series of latch-operated stop buttons to allow the motor to be stopped in an emergency.

Power factor improvement

Power factor has already been discussed in chapter two but it is important to appreciate that a.c. motors, because they have windings, are inductive circuits and as a result will invariably have poor

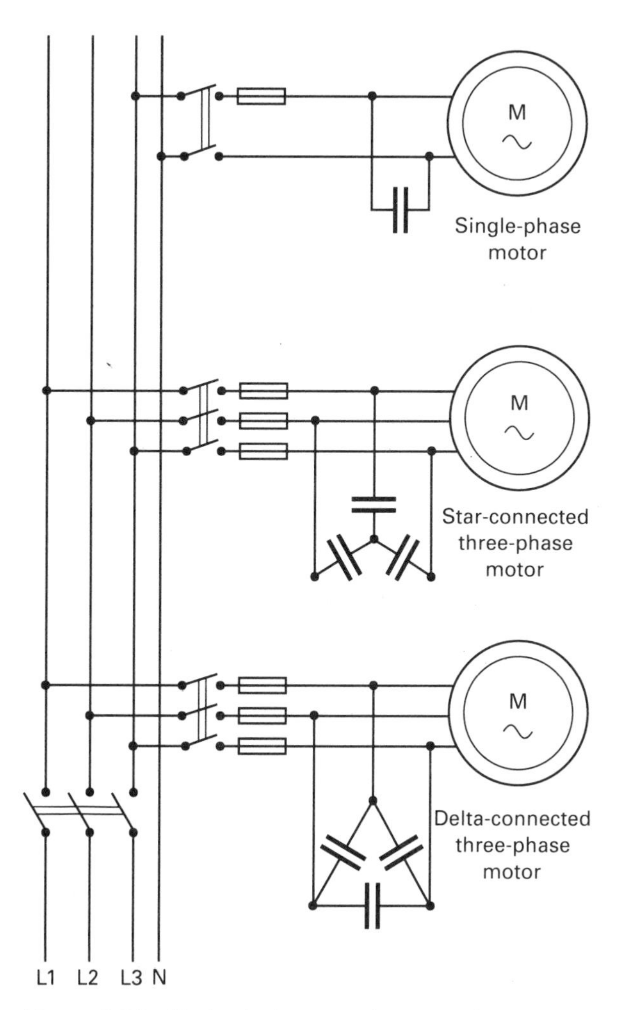

Figure 3.32 Method of connecting capacitors to motors

power factors. A motor at standstill will have a low power factor of about 0.35 but when running on full load it will increase to about 0.7. The most common arrangement to improve power factor is to connect capacitors across motor windings.

In this way it is possible to improve the power factor of the circuit to unity. Figure 3.32 shows several methods depending on a motor's supply connections. In large industrial premises where considerable use is made of motors, automatic power factor correction is installed.

Speed control

From expression [3.9], you will see that the speed of a cage induction motor depends on the frequency of supply and number of pairs of poles in its stator. Methods are available with electronic

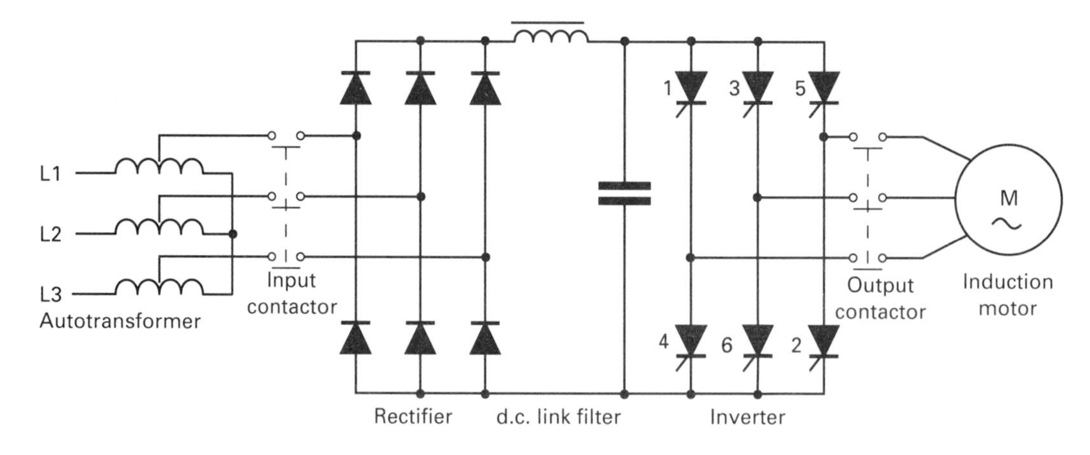

Figure 3.33 Basic inverter system controlling speed of a three-phase induction motor

equipment, such as a static thyristor inverter, tovary the supply frequency over a wide range and thus provide a variable speed output. Figure 3.33 is a simplified circuit diagram of an inverter system controlling a three-phase induction motor. The method of pole-changing in a motor's stator winding is limited to two or three alternative running speeds. In the wound rotor slip-ring motor only a degree of speed control can be achieved with external starting resistance. At low speeds, a motor's efficiency is greatly reduced.

For small motors connected to single-phase supplies, speed control is achieved through the trigger action of a thyristor's gate (see Figure 3.34). In this diagram the two resistors act as a voltage divider. Diode (D2) only allows positive pulses to the thyristor's gate while diode (D1) prevents any feedback from the negative half cycle of the supply. This circuit only provides a limited amount of speed control. By closing the switch the motor can run at its maximum speed.

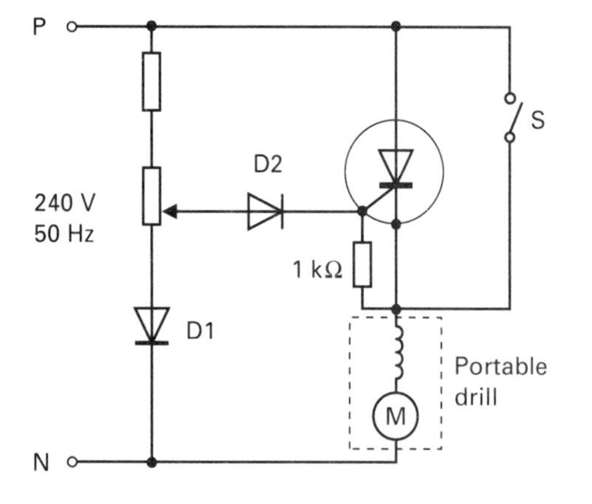

Figure 3.34 Thyristor motor speed control

EXERCISE 3

1. Figure 3.35 shows a simplified diagram of the internal connections of an a.c. induction motor's three-phase stator windings. Using the cross and dot notation of current going in and coming out of a conductor, show on the two adjacent diagrams:
   a) the polarity of the poles for phase positions 1 and 2 and
   b) state the direction of the rotating magnetic field.

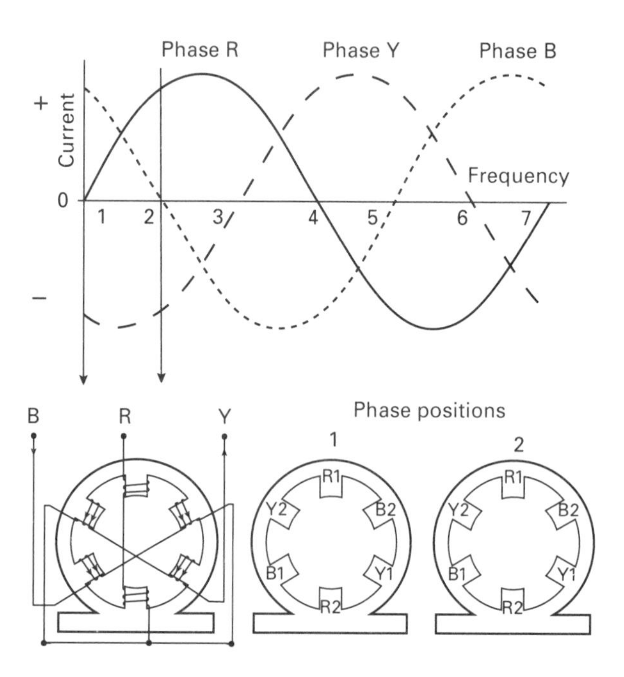

Figure 3.35 Production of rotating magnetic field

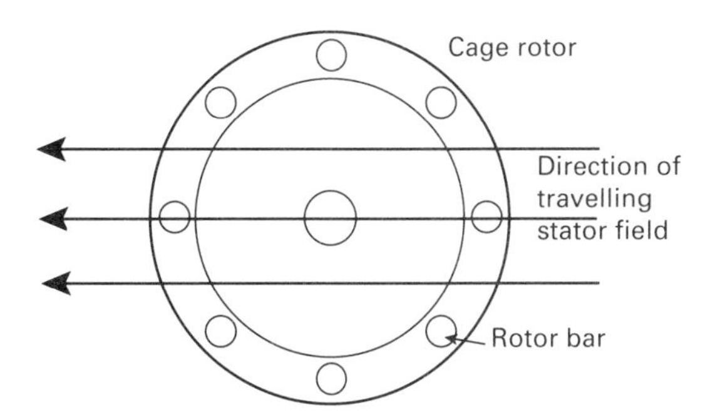

Figure 3.36 Cage rotor end ring

2. a) Determine the synchronous speed and rotor speed of an induction motor for the following conditions:
   (i) $f = 50, p = 2, s = 0.03$
   (ii) $f = 50, p = 4, s = 0.03$
   (iii) $f = 60, p = 6, s = 0.05$

   b) What is meant by the fault termed *single-phasing*?

3. a) Figure 3.36 shows the short-circuited end ring of cage rotor for an induction motor. Using Fleming's right-hand rule, sketch on the rotor the direction of induced current in each rotor bar conductor and show the resultant magnetic field produced.

   b) If the travelling stator magnetic field moves anticlockwise, in which direction will the rotor cage travel?

   c) How would you reverse the direction of a capacitor-start induction motor?

4. a) State ONE advantage and ONE disadvantage of a slip-ring motor over a cage induction motor.

   b) How are remote stop and start buttons connected in a direct-on-line contactor starter?

   c) What is the function of a no-volt release coil?

5. A 1 kW/240 V, 50 Hz 2-pole single-phase induction motor operates with a 5% slip. Its efficiency is 75% and it has a power factor of 0.7 lagging on full load. For this motor:

   a) Draw a labelled circuit diagram of a push button starter with undervoltage protection and overload protection.

   b) Calculate the input power, current and speed.

6. Determine the power output, power factor and efficiency of a three-phase motor having the following test data:

speed	23.75 rev/s
wattmeter	16.92 kW
voltmeter	400 V
ammeter	39.5 A
pulley dia.	0.33 m
brake force	564.44 N

7. Show with a circuit diagram how capacitors can be connected to single-phase and three-phase induction motors to improve their power factor.

8. A universal motor can be connected to an a.c. or d.c. supply. Explain how this is possible. What are the dangers if the motor becomes lightly loaded? How is reversal of direction achieved?

9. Figure 3.37 (over) shows a schematic diagram of two three-phase motors operating a single speed hoist and traverse of a crane. Follow the control circuit wiring and describe the procedure for raising and lowering the hoist and moving the traverse left and right.

10. Figure 3.38 show a circuit diagram of hand-operated star-delta starter. Complete the control circuit wiring to bring in the main contactor.

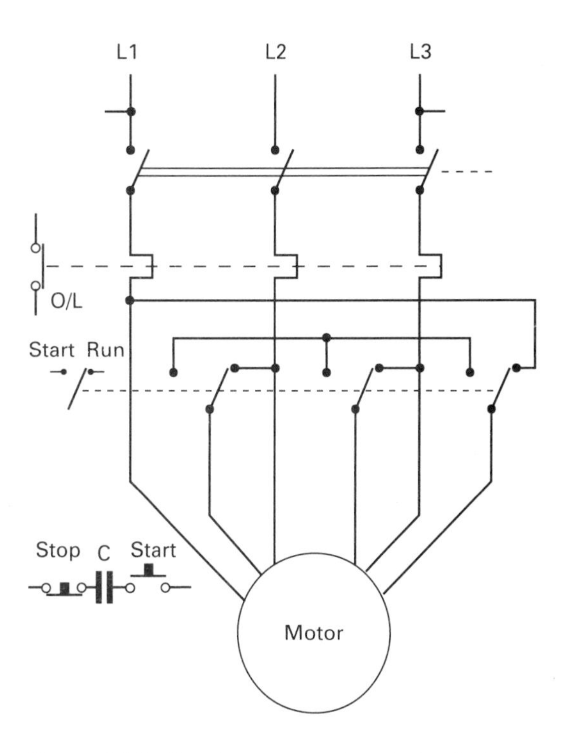

Figure 3.38 Hand operated star-delta starter showing contactor control

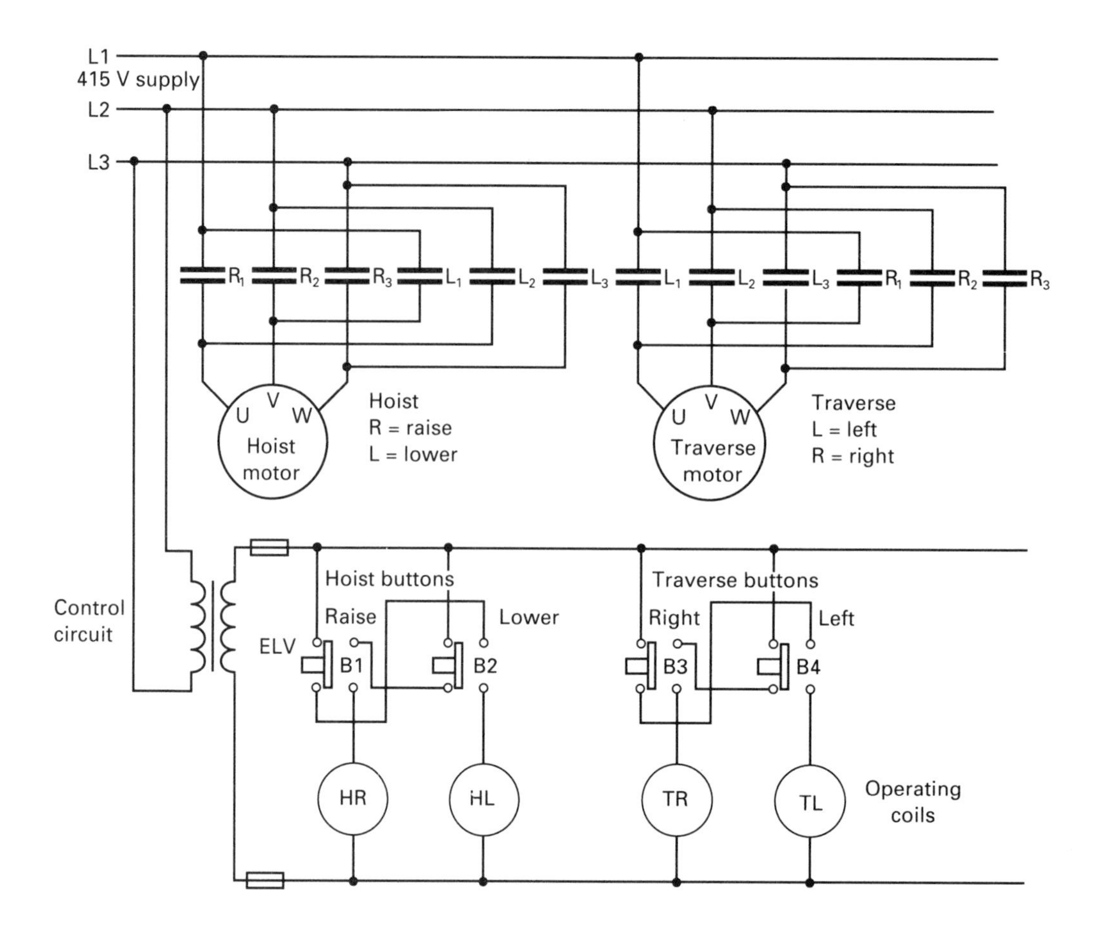

Figure 3.37 Crane control circuit

Elements of lighting design

Objectives

After reading this chapter you should be able to:

- *describe the properties of light in terms of its colours;*
- *state the wavelengths of visible light;*
- *state the meaning of a number of terms relating to lighting design e.g. luminous flux, luminous efficacy, luminous intensity, illumination, illuminance, luminance, inverse square law, cosine law, glare, colour, reflectance, refraction, room index, spacing-height ratio, uniformity, utilisation factor, maintenance factor, lighting design, lumens;*
- *perform simple calculations using the inverse square law and lumen method;*
- *describe general lighting, localised lighting and local lighting systems;*
- *describe various factors which influence the choice of lamp and luminaire for a particular application;*
- *state methods of starting discharge lamps;*
- *state possible faults in various types of lighting circuits.*

Light

Light is a form of energy that we detect with our eyes and is transmitted by electromagnetic waves. Figure 4.1 shows that it occupies only a small portion of the **electromagnetic radiation spectrum**. You will see from the diagram that cosmic rays are at one end and radio waves at the other.

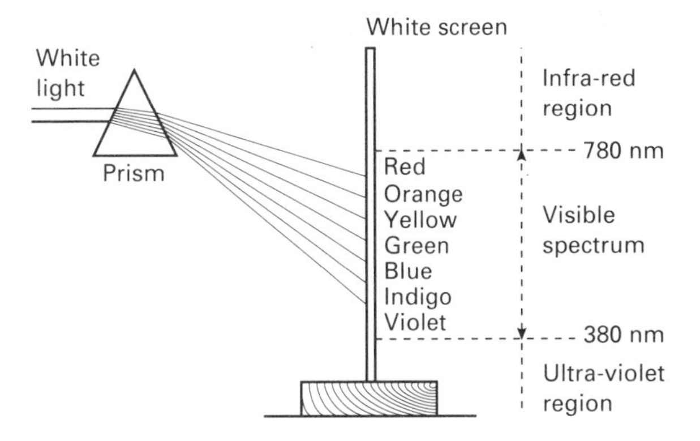

Figure 4.2 Visible light spectrum

Cosmic waves have shorter wavelengths than radio waves but we cannot see them. Visible light lies between 380 and 780 nanometres which is 10^{-9} m. Although it is a form of wave motion, it travels in straight lines at a speed of 300 million metres per second or 3×10^8 m/s. The light from the sun takes eight minutes to reach us and is usually considered to be white light but in actual fact when it is passed through a glass prism it is found to be made up of all the colours of the rainbow. The red colour is refracted the least and the violet colour refracted the most (see Figure 4.2).

The colours have different frequencies, the higher the frequency the shorter will be the wave-

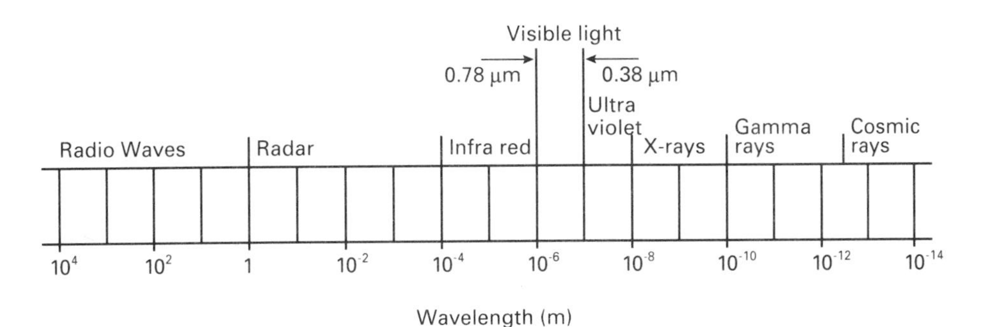

Figure 4.1 Electromagnetic radiation spectrum

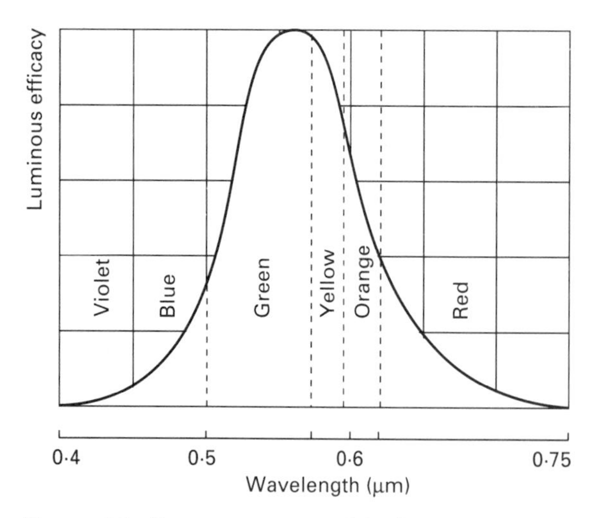

Figure 4.3 Response curve of the human eye

length. The relationship between wavelength (λ), frequency (f) and velocity (v) is given by the following expression:

$$f = \frac{v}{\lambda} \qquad [4.1]$$

Where the symbol λ is the Greek letter *lambda* standing for wavelength. If we wish to find the frequency of, say, the colour 'violet' at 400 nm, from the above expression it is found to be:

$$f = \frac{3 \times 10^{8}}{400 \times 10^{-9}}$$

$$= 7.5 \times 10^{14} \text{ Hz}$$

Figure 4.3 shows the response curve of the human eye. It will be seen that its sensitivity is highest in the middle of the curve, (i.e. in the green and yellow wavebands). Outside these regions the curve tails off into the infra-red and ultra-violet zones. Not only are both extremities present in natural daylight but they are also present in artificial light sources. The incandescent lamp (GLS lamp) mostly emits infra-red radiation in the form of heat. Its spectral distribution curve is shown in Figure 4.4 where it can be seen that its light is mostly at the red end, spreading across the colour band and sloping downwards to the left hand side to produce only a small amount of ultra-violet light. In a fluorescent lamp (MCF lamp), by contrast, the emission is mostly ultra-violet radiation which is converted into visible light by the introduction of various fluorescent phosphors which are coated on the inside of the tube. Figure 4.4 shows a high-pressure mercury vapour discharge lamp. A third lamp worth mentioning is the low pressure sodium discharge lamp (SOX lamp). Its spectral distribution concentrates around 589 nm giving a distinct yellow appearance, making all surrounding colours look yellow.

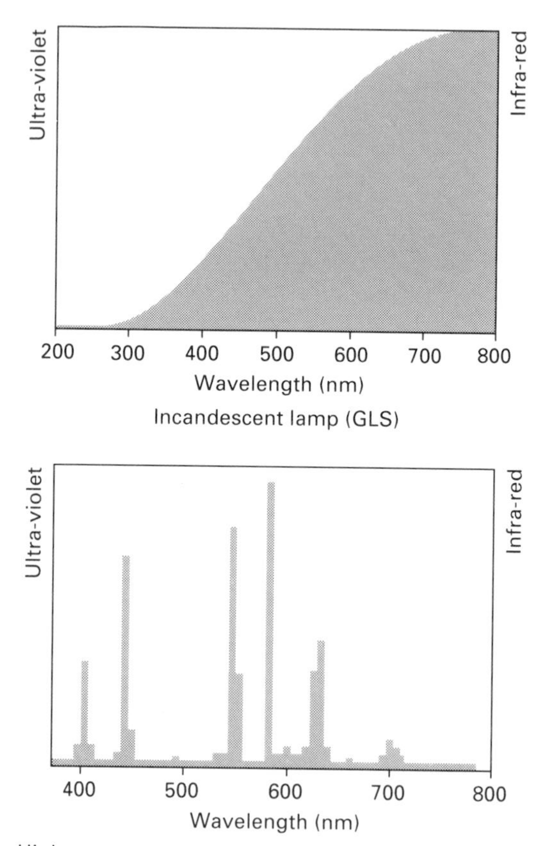

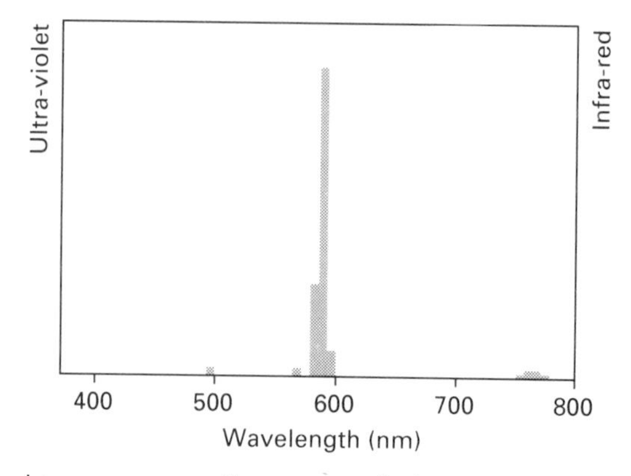

Low-pressure sodium vapour discharge lamp (SOX)

Figure 4.4 Spectral distribution of various lamps

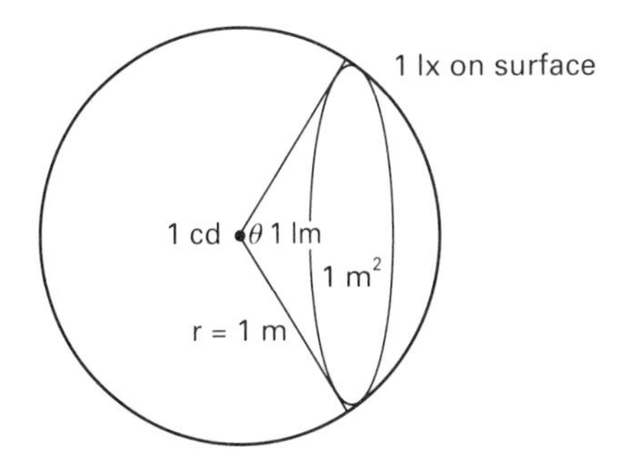

Figure 4.5 Relationships of light units in a solid angle

Terminology

The following are some useful terms used in lighting design:

Luminous flux (F): the rate of flow of luminous energy measured in lumens (lm).
Luminous efficacy: the quantity of light emitted by a source in **lumens** divided by the power source input in watts (lm/W).
Luminous intensity (I): the illuminating power of a light source measured in candela (cd). If a light source has an intensity of 1 **candela**, then 1 lumen will be emitted on 1 square metre of surface, giving an illuminance of 1 lux. In Figure 4.5 since the surface area of a sphere is $4\pi r^2$ and the radius is 1 m, then 4π lumens will be emitted by 1 candela.
Illumination: is the general term for lighting, describing for example a system of leisure and entertainment lighting.
Illuminance (E): is the measure of light falling on a surface measured in lumens per square metre which is the **lux** (lx). More technically it is the luminous flux incident upon a small element of the surface, divided by the area of the element (i.e. $E = F/A$). Some recommended values of illuminance are given in Figure 4.6.
Luminance (L): the measured brightness of a surface measured in candela per square metre which is the nit (nt). It will depend on a number of factors such as adaptation, i.e. the eyes adjustment to light intensity.
Inverse square law: A law which considers the fact that the illumination received on a surface due to a point source is inversely proportional to the square of its perpendicular distance from the source (see Figure 4.7(a).

The illuminance (E) is found by the expression:

$$E = \frac{I}{d^2} \text{ lux} \qquad [4.2]$$

Cosine law: A law which considers the case where the surface to be illuminated is not perpendicular to the direction of the light source. In other words, if the surface is turned so that the rays of light hit it at an angle the illuminated area will increase in size but the illumination will decrease (see Figure 4.7(b)). The illuminance (E) is found by the expression:

$$E = \frac{I \times \cos \phi}{d^2} \text{ lux} \qquad [4.3]$$

Note: The distance (d) in both laws is not the same measurement. In the cosine law the distance can be thought of as the hypotenuse side of a right-angle triangle (h). The adjacent side being the distance (d) if using the inverse square law expression.

Lux level	*Area of activity*
50	Walkways, cable tunnels
100	Bulk stores, corridors
150	Churches Loading bays
200	Monitoring automatic processes in manufacture, turbine halls
300	Lecture theatres, packing goods, rough sawing
500	General offices, kitchens, laboratories
750	Drawing offices, ceramic
1000	Electronic component assembly, gauge and tool rooms, retouching paintwork
1500	Inspection of graphic reproduction, hand tailoring, watch repairs
2000	Assembly of minute mechanisms, jewellers/antiques goldsmith

Figure 4.6 Recommended lux values for different areas of activity

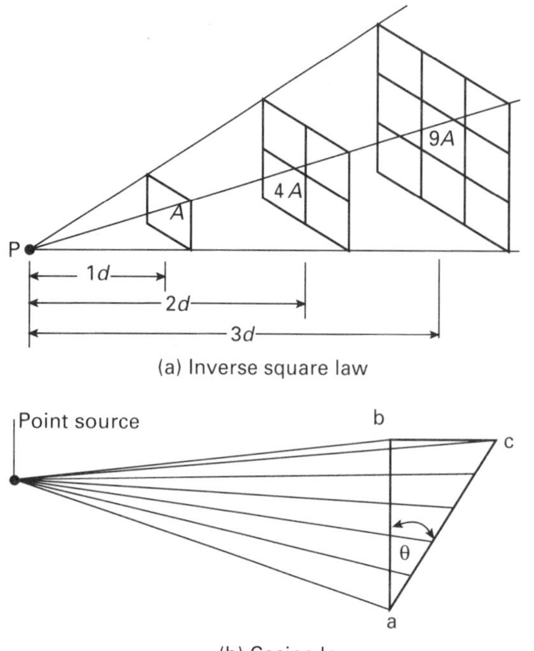

Figure 4.7 Laws of illumination

Q

Example 4.1

An incandescent lamp is suspended 3 m above a level workbench and is fitted with a reflector such that the luminous intensity in all directions below the horizontal is 400 cd. Calculate the illuminance at point A on the surface of the bench directly below the lamp and also at a point B, 4 m away from point A (see Figure 4.8).

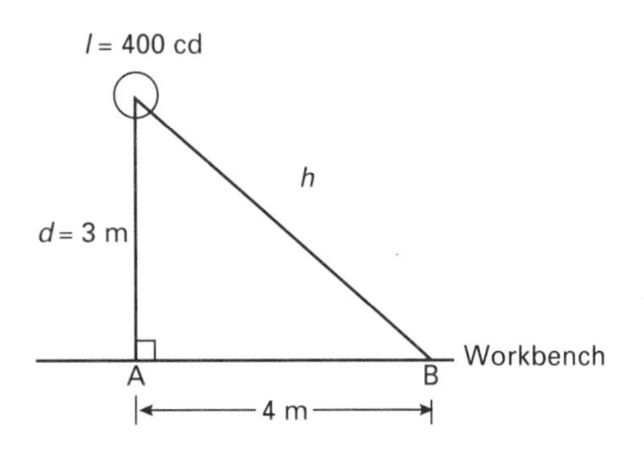

Figure 4.8 Finding the illuminance at two different positions

Solution

Figure 4.8 is a 3–4–5 right-angle triangle. The illuminance at A is found from the inverse square law:

$$E = \frac{I}{d^2} = \frac{400}{9} = 44.4 \text{ lx}$$

The illumination at B is found by using the cosine law, i.e.

$$E = \frac{I \times \cos \phi}{h^2}$$

$$= \frac{400 \times 0.6}{25} = 9.6 \text{ lx}$$

Colour: the spectral distribution of a light source is often a mix of colours, e.g. red, green and blue and are called in lighting **primary colours**. The eye cannot differentiate between mixtures of light giving the same appearance. The term **colour appearance** is used to describe what colour a light source or object appears to be or looks like. The colour of a light source to reveal the colours of an object is called **colour rendering**.

Only those colours which fall onto a surface can be reflected from it. If the light source is, say, deficient in the red wavelength the surface will lack that colour.

Reflectance: a measure of how effectively a surface will reflect light. It is often described as the ratio of luminous flux reflected to the luminous flux received.

Glare: the discomfort or impairment of vision experienced when parts of the visual field are excessively bright in relation to their surroundings. The term 'glare' has several interpretations. It can occur from a reflected surface or directly from a light source. It is often divided into **disability glare** and **discomfort glare**. The latter refers to the degree of visual discomfort that can be tolerated and is often expressed as an index in lighting schemes. Glare can be reduced by using prismatic attachments on luminaires. They are designed to redirect the light of a lamp into its required distribution.

Refraction: is used to describe the bending of light waves when passed through glass or any transparent medium.

Room index: a measure of the dimensions of a room used for calculating the utilisation of a lighting scheme.

For a rectangular room it is given by the expression:

$$\text{Room index} = \frac{\text{Length} \times \text{Width}}{(\text{Length} + \text{Width}) \times \text{Height}} \quad [4.4]$$

The height (H m) is the distance between the luminaire plane and horizontal reference plane (see Figure 4.9). Room indices are given in lighting tables.

Spacing to height ratio: the spacing between luminaires divided by their height above the horizontal reference plane.

Working plane: the horizontal, vertical or inclined plane in which the visual task lies.

Uniformity: the ratio between the minimum and average illuminance over the working area. It should not be less than 0.8 where tasks are performed.

Utilization factor: the total flux reaching the working plane divided by the total lamp flux. Its value will depend on room index and effective reflectances of standard surfaces, e.g. floor cavity (F), walls (W) and ceiling cavity (C). Specific values are often given in lighting tables.

Maintenance factor: the ratio of illumination from a dirty installation to that from the same installation when clean, expressed as a decimal (often 0.8). It is often referred to as the **loss life factor** (LLF) being the product of four other factors, namely, luminaire dirt depreciation (LDD), lamp failure factor (LLF), lamp lumen depreciation (LLD) and room dirt depreciation (RDD).

Lighting design lumens (LDL): the average lumen outputs of lamps throughout their life, based on 2000 hours of use. They are less than the initial lumen values when the lamps are new.

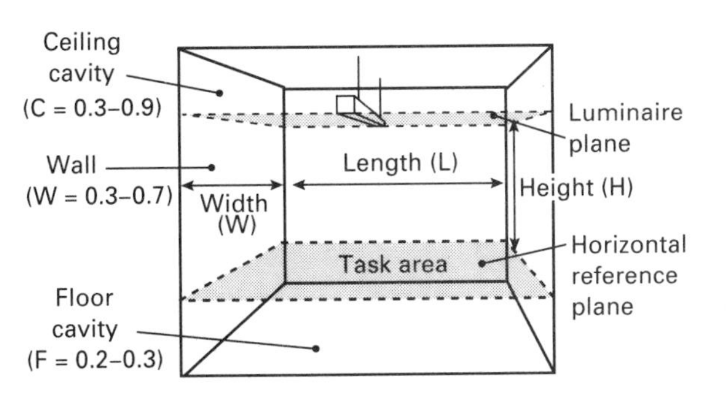

Figure 4.9 Method of finding room index

Lumen method calculations

This is the recognised method of finding the total lumens required of a lighting installation. It is expressed in the form of average illuminance:

$$E = \frac{F \times n \times N \times LLF \times UF}{A} \text{ lux} \quad [4.5]$$

Where: E is the bare lamp flux per lamp
n is the number of lamps per luminaire
N is the number of luminaires
LLF is the light loss factor for a surface
UF is the utilisation factor for the surface
A is the area of the surface

By transposition of formula [4.5]:

$$F = \frac{E \times A}{n \times N \times LLF \times UF} \text{ lumens}$$

A simple example of this method is shown below.

Q

Example 4.2

A rectangular-shaped workshop building 25 m long and 40 m wide is to be illuminated using 60 single-tube 70 W fluorescent luminaires. Each tube has an output of 6300 lumens and the input to each luminaire is 82 W.

Determine:

a) the average illuminance if the light loss factor is 0.8 and the utilisation factor is 0.66.
b) the efficacy of the luminaires.

A

Solution

a) $$E = \frac{F \times N \times LLF \times UF}{A} = \frac{6300 \times 60 \times 0.8 \times 0.66}{25 \times 40} = 200 \text{ lux}$$

b) Efficacy = lumens/watts
= 6300/82 = 77 lm/W

Q

Example 4.3

A classroom measuring 6 m × 14 m requires to be illuminated to 500 lux by single-tube fluorescent lamps having a lumen output of 4408 lumens. Each lamp is rated at 58 W and the light loss factor and utilisation factor are 0.9 and 0.5 respectively.

a) How many luminaires are required to give the average illuminance?
b) If the supply is 415 V three-phase, four-wire, how many luminaires can be connected to each phase for load balancing?
c) Calculate the current per phase, if each luminaire has control gear losses of 12 W.
d) What is the efficacy of each fluorescent tube and each luminaire?

A

Solution

a) $N = \dfrac{E \times A}{F \times LLF \times UF}$

$= \dfrac{500 \times 6 \times 14}{4408 \times 0.9 \times 0.5} = 21 \text{ lamps}$

b) 7 luminaires per phase.

c) $I = \dfrac{P \times 1.8}{V}$

$= \dfrac{7 \times (58 + 12) \times 1.8}{240}$

$= 3.68 \text{ A}$

d) Efficacy of tube in lamp:

lumen/watt = 4408/58

= 76 lm/W

Efficacy of luminaire is:

lumen/watt = 4408/70

= 63 lm/W

Note: In (c) above, since the power factor of the circuit is not given the multiplier 1.8 has been used and assumes a p.f. of not less than 0.85.

Example 4.4

A lighting installation in a room is to provide an average illuminance of 500 lx on the working plane. The dimensions of the room are: length 30 m, width 7.5 m and height 3.15 m. The working plane is 0.75 m. Determine the total luminous flux and the number of twin fluorescent luminaires required for the installation if the light loss factor is 0.8 and the utilisation factor is 0.71.

A

Solution

With the above information the formula is written for the total lumens required:

$F = \dfrac{E \times A}{LLF \times UF}$

$= \dfrac{500 \times (30 \times 7.5)}{0.8 \times 0.71}$

$= \dfrac{112\,500}{0.57} = 197\,368 \text{ lm}$

If the selected lamps were given lighting design lumens (LDL) of 4450 per luminaire then the number of lamps required would be:

$N = \dfrac{197\,368}{4450} = 44$

If twin luminaires were chosen, then only 22 luminaires would be needed. In order to extend this example, assume the room had the following reflectances: floor cavity (F) = 0.2, wall (W) = 0.5 and ceiling cavity = 0.7. To find the utilisation factor, we also need to know the room index (RI). This is derived from:

$RI = \dfrac{\text{Length} \times \text{Width}}{(\text{Length} + \text{Width}) \times \text{Height}}$

$= \dfrac{30 \times 7.5}{(30 + 7.5) \times 2.4} = 2.5$

With the reflectance indices and the room index the utilisation factor can easily be found from lamp data tables (e.g. UF = 0.71 for a *'Thorn' Clipper 2* luminaire).

If each fluorescent lamp is rated at 58 W, its efficacy would be 4450/58 = 77 lm/W.

Lamp data tables provide the spacing to height ratios (SHR) for luminaires, quoting the maximum (SHR MAX) and nominal (SHR NOM) spacings in metres. For planning lighting on ceilings, it is appropriate to consider the transverse (TR) and axial (AX) spacings between luminaires and the luminaires and side walls. In the above example, from lighting tables, the chosen luminaires to produce a uniform illuminance must conform to SHR MAX of 1.74 m, SHR TR MAX of 2.02 m and a

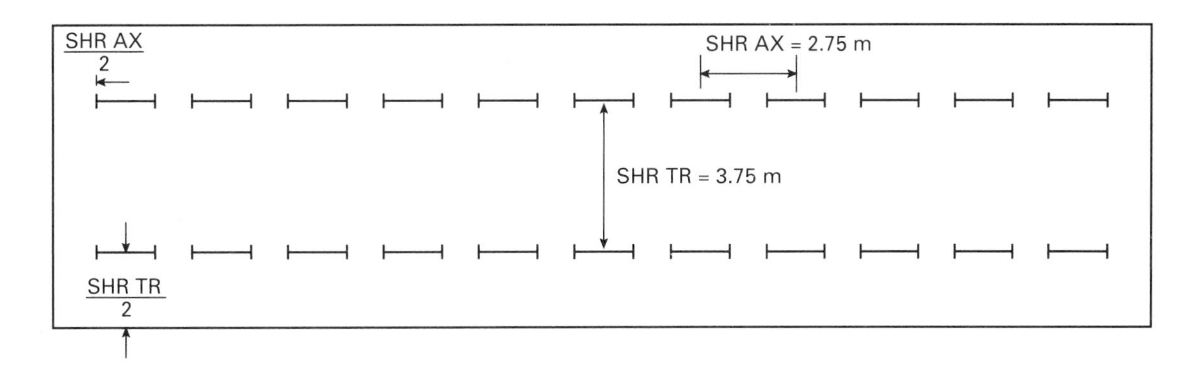

Figure 4.10 Lighting layout

SHR NOM of 1.5 m. Since H = 2.4 m, then SHR AX NOM is $1.5 \times 2.4 = 3.6$ m but should not exceed 4.2 m. Also, SHR TR MAX is $2.02 \times 2.4 = 4.85$ m. With this information the layout of the luminaires should comfortably fit into two rows of eleven luminaires per row, i.e.

Length:

$$(2 \times 0.62 \text{ m}) + (11 \times 1.5 \text{ m}) + (10 \times 1.23 \text{ m}) = 30 \text{ m}.$$

Width: $(2 \times 1.87 \text{ m}) + 3.75 \text{ m} = 7.5 \text{ m}$

Figure 4.10 is a diagram of the layout.

The above example has been made simple to explain the basic procedure involved in the lumen method calculation. In practice, several other factors may need to be considered in the layout, not least the task area to be illuminated and individual rooms.

Lighting systems

Lighting systems can be classified as:

a) general lighting
b) localised lighting and
c) local lighting (see Figure 4.11).

General lighting is mostly seen as an arrangement of ceiling luminaires intended to provide approximate uniform illuminance over the horizontal working plane. The design is based on the lumen method of calculation and lends itself to freely changing work patterns of non-critical tasks. Nowadays, it is directed at the comfort levels of room occupants with emphasis placed on creating a natural environment to work in, minimising problems of glare. This is a particular problem in offices where a large number of VDUs (visual display units) are used. The problem is often overcome by using luminaires with opal plastic **diffusers** or by installing luminaires recessed within suspended ceilings. Prismatic controllers or low-brightness plastic parabolic **louvres** which direct the light straight downwards provide a high degree of glare control (see Fig 4.12).

It could also be achieved by using a system of

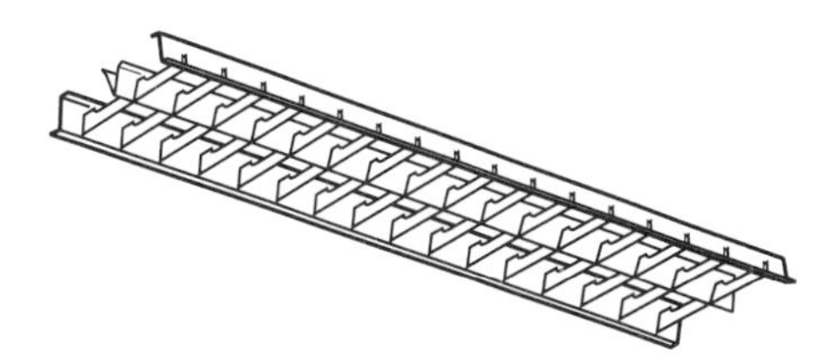

Figure 4.12 Low-brightness "batwing" louvres

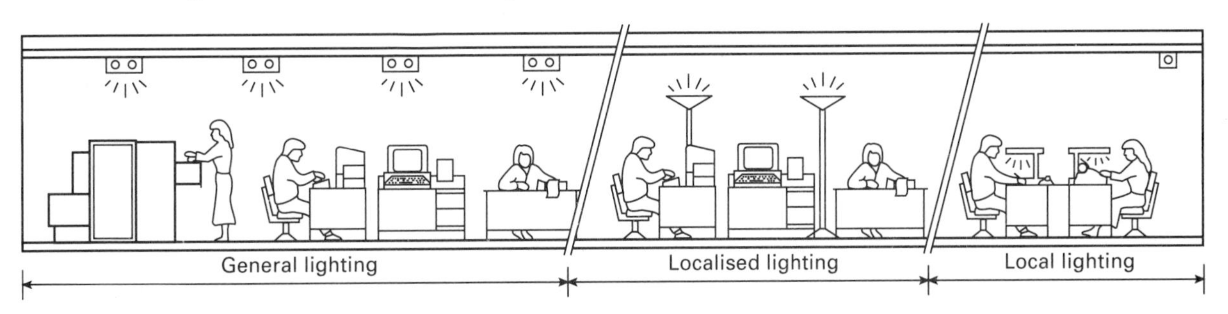

Figure 4.11 Interior lighting schemes

uplighting which is a form of **indirect lighting** reflected from the ceiling. The luminaires can either be floor standing, wall mounted or suspended types and are best installed within the system's furniture hidden from view (see Figure 4.13).

In **localised lighting**, careful positioning of the luminaires is of prime consideration in order to reduce glare and shadow. The method has to provide the necessary illuminance over the required working area and give only about 50% illuminance to the immediate surroundings. It finds considerable use in supermarket stores for display lighting over rows of food products and it can be designed to use less energy than general lighting by offering independent control.

Local lighting can be regarded as **task lighting**, providing the necessary illuminance over small areas, such as desks, machines, display areas etc. The luminaires chosen must provide a higher illuminance than those used for general lighting (about three times as much) and they can incorporate adjustment facilities to avoid any possibility of glare. The use of **spotlights** for display purposes and for special effects offers personal control of the area being illuminated. It should be seen as being quite independent from general lighting.

Choice of lamps and luminaires

It is said that a good lighting scheme can raise the quality of life, not only by improving the environment in terms of comfort and pleasure but also by increasing productivity, reducing accidents, improving security and reducing maintenance costs.

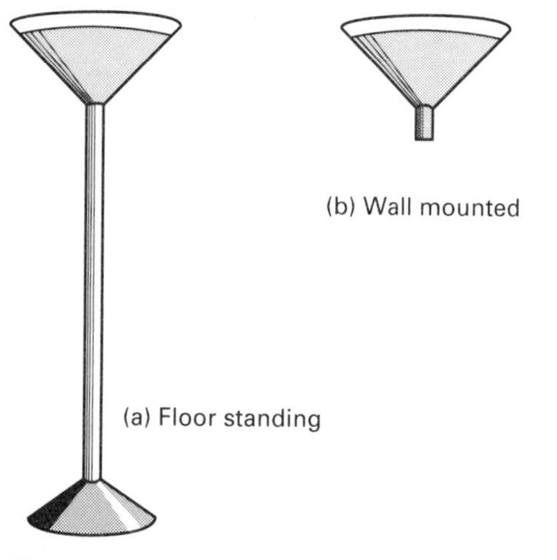

Figure 4.13 Up-lights

Different types of lamp in general use are covered in Part I Studies, Theory book, by the same author. This explained the operation and characteristics of several types of lamp together with wiring diagrams of their control circuits.

A lighting designer will almost certainly consider the following points when deciding a type of lamp for a particular lighting scheme:

1) colour rendering (the lamp's colour quality);
2) colour appearance (the lamp's apparent colour based on its correlated colour temperature – warm, intermediate, cool or cold);
3) knowledge of the run-up time to full brightness and also its re-strike time;
4) suitability, efficacy, life, and need for starting control gear;
5) protection from substances in the environment (internal or external);
6) energy costs and lamp maintenance/replacement costs.

A **luminaire** is a combination of lamp and lighting fittings and its purpose is to redirect the light from the lamp in a particular direction, preferably with the least glare possible. A classification for general interior lighting considers the total light output of a luminaire in the upper and lower hemispheres. For example, **direct lighting** is regarded as producing 90–100% light shining downwards and only 0–10% light shining upwards, whereas **semi-direct lighting** produces 10–40% light upwards and 60–90% light downwards. The percentages for **indirect lighting** and **semi-indirect lighting** are opposite to the direct and semi-direct percentages.

Figure 4.14 illustrates these methods together with **general diffuse lighting** which produces between 40–60% upwards and downwards.

A major factor in selection, apart from cost, is the luminaire's protection properties. It may need mechanical protection to protect the lamp against vibration or vandalism or it may be designed for

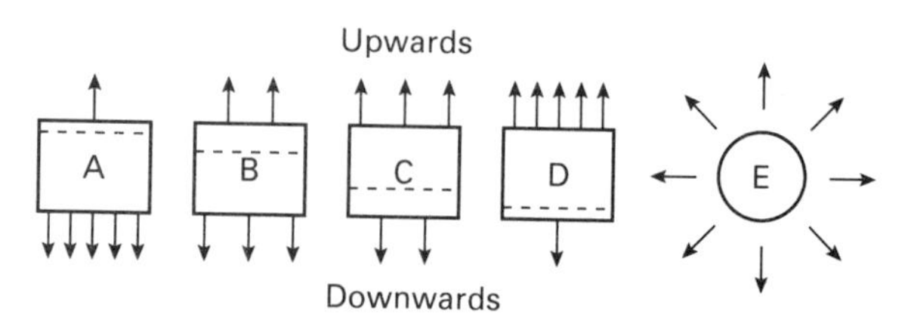

A – direct; B – semi-direct; C – semi-indirect; D – indirect; E – general diffusing

Figure 4.14 Luminaire classifications

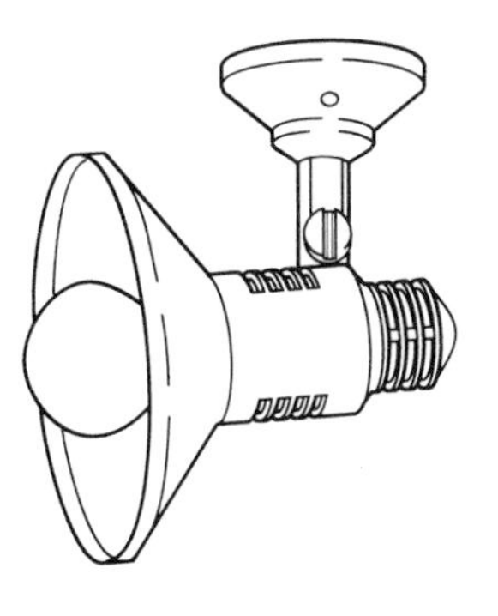

Lux (Beam angle 80°)			
Metres	*60w*	*75w*	*100w*
1	270	360	530
2	70	90	135
3	30	40	60
4	17	23	35

Figure 4.15 Beam spotlight luminaire

strength of support fixing. It may also need protection against corrosion or explosion, heat, dust, water or chemical contamination. Some luminaires may be designed with a reflector to obtain a brighter beam.

Figure 4.15 shows a bowl reflector spotlight suitable for commercial interior display lighting often used in hotels, shops and galleries, etc. The body of the luminaire is heat-resistant and flame-retardant and incorporates a 100 W, Crown Silvered, ES lamp which gives high performance and minimum glare. At the side of the luminaire is a beam cone diagram giving the lux values based on the inverse square law calculation. These values have been taken at the beam centre for the distances shown.

Types of lamp

Common lamps and their control circuits are discussed in *Part 1 Studies: Theory* by the same author. Only a summary of their operation will be mentioned here. It is generally accepted that lamps for artificial lighting can be classified according to the way they produce light. For example, the **tungsten filament lamp (GLS)** and **tungsten halogen lamp (TH)** produce their light from incandescence, that is, light from a heated filament. Tungsten is an ideal material for the filament because of its high melting point (3380°C). In the GLS lamp, gas filling is used to improve life expectancy and will be either argon or nitrogen to stop filament evaporation from the high internal temperature. In some projector lamps, a blend of both gases is used but in low wattage lamps below 25 W no gas filling is required. The actual bulb of the lamp can be made 'clear' or it can be internally 'etched' to give a pearl finish to reduce glare and soften shadows. Coiling a lamp's filament increases both life expectancy and efficacy and double life lamps of 2000 hours are quite common. GLS lamps are relatively cheap and various shapes, colours and lamp caps exist. Among the range in use today can be found rough service lamps used where shock and vibration are unavoidable and reflector lamps for spotlighting and floodlighting application. The pressed glass lamp, known as the PAR lamp (parabolic anodised reflector) has its filament placed in the focus of the reflector so as to provide a directional beam.

In the TH lamp, the bulb is made of quartz (a very hard transparent mineral) instead of glass. This means that it can safely operate at high temperature. This allows the internal gas pressure of the halogen to be raised, reducing filament evaporation and by doing this the average life of the lamp and its efficacy can be considerably increased (a 100 W TH lamp has a life expectancy of around 4000 hours). With a much improved light output and good colour rendering properties, as well as warm appearance, the TH lamp is ideal for commercial use such as display lighting and exterior floodlighting as well as specialised automobile lighting.

A second type of artificial light source is the **fluorescent lamp**, called a low pressure mercury vapour lamp (MCF). The lamp is a form of two-action process light source, since ultra-violet radiation emitted from the internal discharge is converted into visible light by the use of phosphors coated on the inside of the tube. The lamp is filled with argon and krypton gas as well as small amounts of liquid mercury. By mixing various phosphors (halophosphates) a wide choice of colour is available. You will come across fluorescent tubes called: white, warm white, cool white (daylight), natural, kolor-rite (tru-colour), northlight (colour matching) and artificial daylight.

Some of these lamps have been superseded by the development of **triphosphors**, now in 'pluslux' lamps and 'polylux' lamps. These lamps produce good colour rendering to give coloured objects a natural look. Their high light output is in the red, green and blue wavelengths and they are designed to save energy. Standard T8 krypton-filled lamps are used in all new installations with T12 argon-filled lamps a second choice in older installations, mainly for lamp replacement. The life and lumen depreciation of fluorescent lamps is, in general, extremely good with triphosphor lamps losing only 90% of their light output after 12 000 h of use.

Another range of smaller fluorescent lamps slowly replacing GLS lamps because of their energy saving costs are **fluorescent compact lamps**. Power consumption of the 2D and SL lamps is about 25% of the GLS lamp and the average life is about 5–10 times longer. With no starting control gear needed an 'SL9W' lamp with BC cap is equivalent to a 40 W GLS lamp and an 'SL25W' lamp is equivalent to a 100 W GLS lamp. These type of fluorescent lamps produce good colour rendering using triphosphors. Today there are several different versions of the compact fluorescent. One type, called the PL 'cluster' lamp has a built-in electronic quick start unit and is fitted with standard BC or ES cap.

Over recent years there have been major advances in **lamp starting** methods. Switchstart circuits using glow-type and electronic-type starters are still widely used and while relatively cheap, suffer a number of minor problems. In glow-type starters the internal bi-metal contacts often stick and it leads to problems of lamp flicker and the eventual overheating of the choke. Today, the starting of fluorescent tubes is often achieved using high frequency electronic ballasts which operate between 20 kHz and 40 kHz. Not only does this raise the efficacy of the luminaire by reducing ballast losses but it also reduces energy consumption. There are numerous other advantages, such as no lamp flicker or stroboscopic effect problems; silent starting and silent running operation; easier maintenance with no additional controlgear; improved circuit power factor, longer life expectancy as well as reduced ballast weight and low ambient temperature within the enclosure of the luminaire.

The third type of artificial light source is known as the **discharge lamp**. Several types of lamp are commonly used but basically they obtain their light from the passage of electricity through a gas or vapour, notably sodium or mercury vapour. Nearly all these lamp require control gear. One type, known as the **low pressure sodium vapour lamp** (SOX), is characterised by its monochromatic yellow light as seen from the lamp's spectural distribution in Figure 4.4(c). Because of this it has poor colour rendering properties and is mostly used for lighting highways, car parks, construction sites and security areas. Figure 4.16 shows the construction of this lamp. When it is first switched on the initial discharge occurs through a mixture of neon and argon gas giving a reddish appearance. Sodium globules on the inside of the inner U-shaped arc tube gradually become vapourised and the lamp's colour characteristics change to yellow. The lamp has an outer glass envelope and the space between this and the arc tube is completely evacuated. To keep the discharge at its optimum temperature (270°C) the internal surface of the outer glass envelope is coated with a layer of indium oxide. This oxide allows light to pass through but reflects infra-red radiation back to the arc tube. The efficacy value of this type of lamp is around 170 lm/W. All lamps are fitted with BC lampholders. Lower wattage lamps can operate in any position but the higher wattage lamps are only

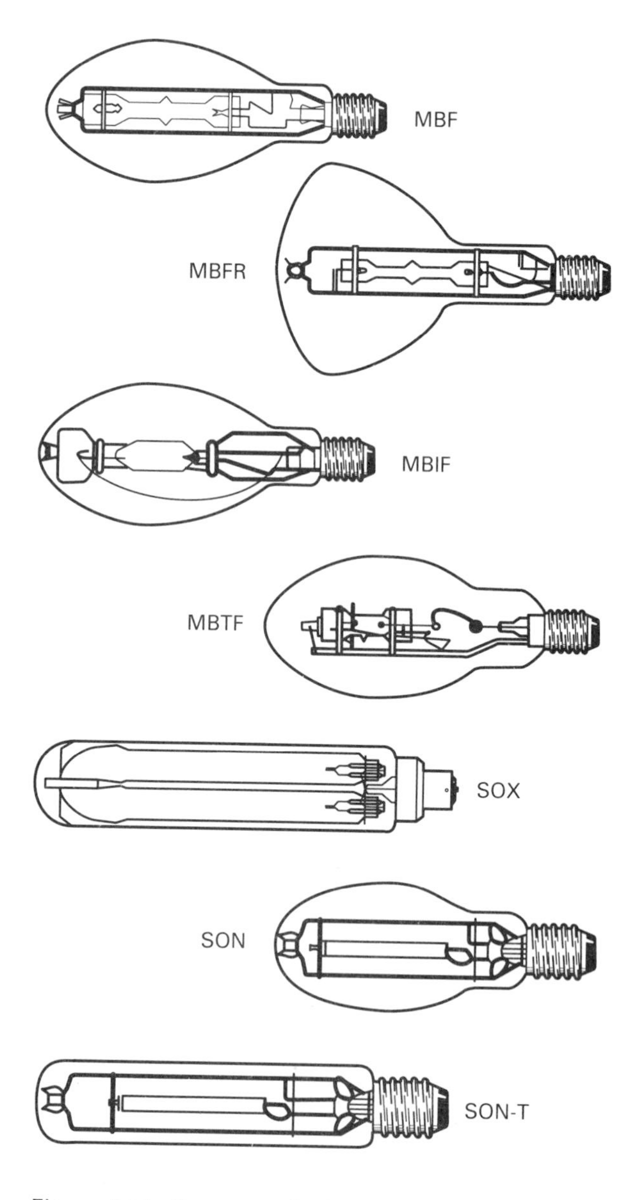

Figure 4.16 Common discharge lamps

allowed to operate up to 20° above or below the horizontal. The control gear for these lamps often requires a 'leaky' reactance transformer ballast or electronic ignitor to assist starting. The sodium lamp takes about 8–9 minutes to reach full brightness and if temporarily switched off its re-ignition time (which may be instant for a low wattage lamp) can take 10 minutes for the higher wattage lamp (90 W and 135 W lamps).

A further development of the SOX lamp is the SOX-E lamp which is more economical having improved thermal insulation and increased efficacy. Lamp wattages between 26 W and 131 W have replaced the SOX lamp standard wattages between 35 W and 180 W. Increasing the internal pressure inside a sodium-vapour lamp by 100 000 times increases its colour rendering.

This improvement has led to the development of the SON lamp which is called a **high pressure sodium vapour lamp**. These lamps operate in a sintered alimina arc tube containing sodium dosed with mercury and argon or xenon as initial starting gases. The arc tube (see Figure 4.16) is sealed into an evacuated outer jacket to reduce power loss and also to prevent oxidation. Once started from high voltage pulses the sodium vapour takes over, reaching full brightness within a few minutes. The light produced is a pleasant golden-white light with warm appearance and adequate colour rendering. Several versions of the lamp are in use today, such as the SON-E, SON-H, SON-R, SON-T, SON-TD, SONDL-E and SONDL-T.

One development of this lamp, known as the **SDW-T lamp** is only 145 mm in length. It is a single-end lamp with an efficacy of 40 lm/W and a life expectancy five times longer than the GLS lamp. It produces a warm white light and finds considerable use in display lighting.

The **SON-E lamp**, has a diffused ellipsoidal outer bulb and is a single-ended lamp. This and the **SON-T** lamp which has a clear tubular shaped outer bulb have wattages ranging between 50 W and 1000 W. They are used mostly in warehouses, for road lighting, floodlighting and security lighting.

The **SON-H lamp** is a self-starting plug-in lamp fitted with a GES cap. It is available in two sizes, 210 W and 350 W and takes about 5 minutes to reach full brightness, producing a golden white light to give fair colour rendering. Its efficacy is about 90 lm/W and it takes 3 minutes to restrike after being temporarily switched off. It finds use in sports halls and public buildings.

SONDL-E and **SONDL-T lamps** operate at higher pressures than ordinary SON lamps and are available in tubular or diffused elliptical shapes. The lamps are rated at 150 W, 250 W and 400 W and have improved colour rendering. They take about 8 minutes to reach full brightness but can restrike in less than 1 minute.

Another type of discharge lamp is the **high pressure mercury vapour lamp** (MBF). Standard MBF lamps are well established and Figure 4.16 shows a typical lamp style. The discharge operates in a quartz arc tube which is coated with a fluorescent phosphor to improve colour rendering. The arc tube contains argon gas and a small quantity of mercury. The initial discharge takes place between an auxiliary electrode and one of the lamp's main electrodes. It then spreads across both main electrodes, taking several minutes to reach full brightness producing a cool white light that provides acceptable colour rendering. Lamp efficacy values range from 38 lm/W to 56 lm/W and wattages range from 50 W to 1000 W and the restrike time is between 4 and 7 minutes. They find considerable use in road and street lighting, public buildings and general amenity lighting. Other lamps of this type are: MBFR (internal reflector), MBFSD (internal phosphors 'Super Deluxe' to give improved colour rendering), and MBI (metal halide).

There are several versions of **metal halide** lamp. The arc tube contains metal halides such as sodium and gallium which have the effect of subduing the mercury spectrum and increasing the light output and colour rendering of the lamp. The lamp is fitted with a GES cap and requires a high starting voltage from an ignitor. It reaches full brightness in 3–5 minutes but its restrike time may take up to 7 minutes. With efficacy values around 90 lm/W and wattages between 250 W and 2 kW typical application of this lamp is in fashion stores, printing rooms, studio lighting and floodlighting.

Lighting control

It is often a requirement in general lighting schemes and special lighting installations such as stage lighting, television rooms, auditoriums, conference rooms, etc, to provide facilities for changing illuminance levels, not simply by the manual switching of rows of luminaires 'on' or 'off' as and when required but by means of manual dimming or automatic control. Dimming is a process of converting the stable a.c. supply into a variable output voltage. In most incandescent lamp circuits it is achieved by varying the potential difference across a resistor connected in series with a capacitor. Increasing the value of the resistance increases the time taken for the capacitor to reach the necessary switching voltage to allow a solid-state device to conduct. A description and diagram of a simple dimmer circuit is given in Appendix 2.5.

In more complex lighting schemes where daylight linking is of prime consideration, one method is to use **photo-cells** or sensors linked with **high frequency regulation ballast** (HFR). This can allow the lighting levels of the installation to be varied between 10% and 100% of full light output.

Lamp circuit faults

Incandescent lamp faults are not too difficult to trace since they do not require control gear. Faults are often due to lamp failure or enclosure damaged or perhaps the lampholder broken or loose connections. A GLS lamp is often near the end of its useful life when the outer glass bulb looks black on top. To avoid the filament shorting out the lead-in wire is fused. In the TH lamp, especially the linear type, vibration or mis-alignment of its contacts may be the reason why it is not working. It is important to read lamp literature because some linear lamps must only be operated within a few degrees of their horizontal mounting position. With this lamp and single-ended lamps, care must be taken to avoid grease (from fingers) touching the outer quartz, otherwise it will crack during operation at the high internal temperature. It is important to see that the wiring insulation directly connecting the lamp is covered with high temperature sleeving.

In the MFC lamp, a number of causes of failure could be attributable to lamp failure. For example, a lamp that fails to strike, showing no signs of life, could have a ruptured fuse, faulty tube or even open circuit choke. If it appears to glow brightly at both ends it is likely to be crossed connections or short-circuit across the lampholder. If it only glows brightly from one end then the problem is in the starter switch. This is also a symptom of a lamp that flashes 'on' and 'off' but it could also mean the end of the lamp's life or even low voltage. If both ends of the lamp are blackened, it is probably the end of its life. Most problems are found by trial and error and the remedy found by replacing components.

With regard to discharge lamps, if a high-pressure mercury lamp refuses to light, it could be a faulty lamp or no supply voltage or even open circuit wiring. It sometimes signifies the end of a lamp's life but it may be caused by insufficient restrike time. If the lamp has a poor light output, it could mean end of life or that it is not connected to the correct ballast tapping. An unstable or flickering light may also mean the end of life approaching. A low pressure sodium lamp which does not light, generally means that the lamp has failed or there is no supply voltage. It may be a problem with the ignitor or with the ballast. Low light output is a sign that its life has been reached. High pressure SON lamps have similar problems.

EXERCISE 4

1. a) Figure 4.17 shows sketch of a GLS lamp. Without reference to any other sources of information, label the components indicated.
b) Explain why a GLS lamp is filled with an inert gas.
c) A coiled-coil GLS lamp of 40 W produces 390 lighting design lumens. What is the lamp's efficacy?

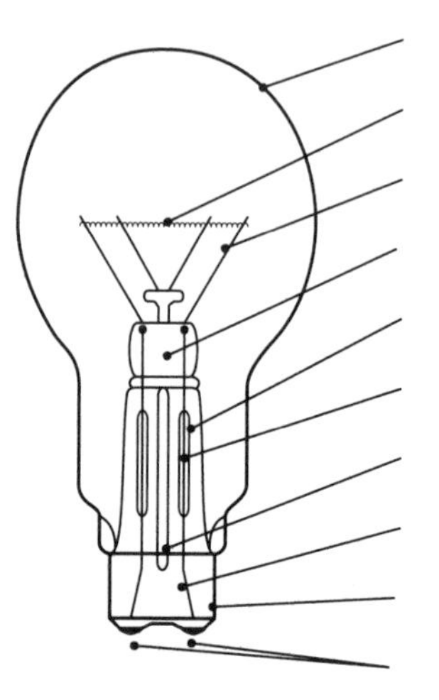

Figure 4.17 GLS lamps

2. a) Briefly describe the following terms:
(i) luminous intensity;
(ii) luminous flux;
(iii) illuminance.
b) A SOX lamp has a poor colour rendering. Why is this?

c) With reference to other sources of information, draw a fully-labelled circuit diagram of the SOX and its control gear.

3. a) An incandescent lamp of luminous intensity of 100 cd in all directions is to provide 40 lux on the surface of a bench directly below the lamp.

Determine the lamp's distance from the surface using the inverse square law method of finding illuminance.

b) If the lamp was lowered by 0.58 m, what would now be the illuminance received on the bench?

4. a) A room measures 10 m × 8 m and requires to illuminate to a level of 500 lx by fluorescent luminaires each having lighting design lumens of 4000 lm. Calculate the total luminous flux and total number of luminaires required if the maintenance factor and utilisation factor are 0.8 and 0.5 respectively.

b) What is meant by the term 'maintenance factor'?

5. a) Describe how you would reduce glare from an open type luminaire.

b) What is the difference between localised lighting and task lighting?

c) State TWO important factors to be considered in lamp selection.

6. Explain with the aid of a circuit diagram the operation of a fluorescent lamp started with a glow-type starter switch.

7. Figure 4.18 shows a high-pressure mercury vapour discharge lamp.

a) What is the purpose of the auxiliary electrode?

b) Why does the lamp refuse to light up immediately after it has been switched off?

8. a) Explain how the discharge occurs in a low-pressure sodium lamp.

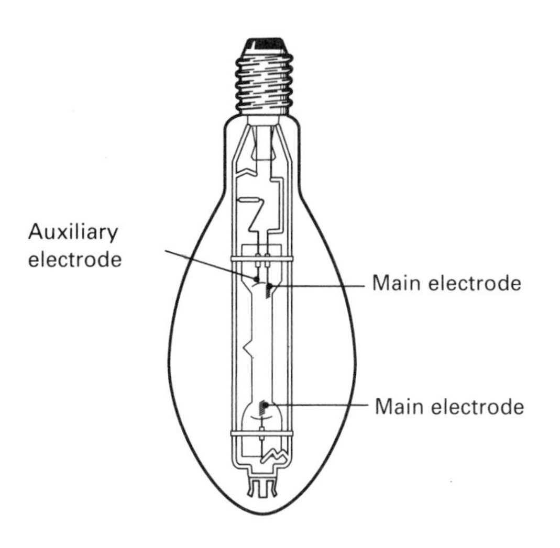

Figure 4.18 MBF discharge lamp

b) What is the burning position of this lamp?

c) State TWO applications.

9. a) Figure 4.19 is a circuit diagram of lamp dimmer circuit. What are the components X and Y?

b) What is the connection called between X and Y?

c) What is the purpose of the resistor R and capacitor C?

10. A high-pressure mercury vapour discharge lamp (MBF) shows symptoms of light fluctuating. State some of the possible causes.

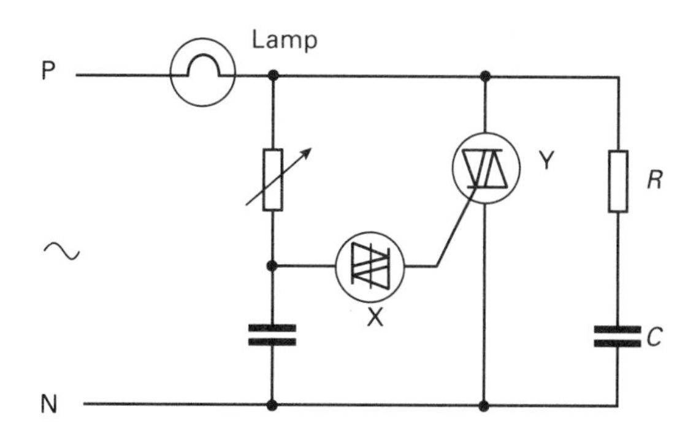

Figure 4.19 Dimmer circuit

Appendix 1

Multiple-choice questions

Objective

After reading this book you should be able to complete the following multiple choice questions.

A multiple-choice question comprises either a direct question or a statement followed by several suggested answers or completions called responses. Only one of these responses is correct, the others are distractors and are aimed at being plausible. In general, four responses are provided and the reader is required to select the letter A, B, C or D which represents the correct response.

INTRODUCTORY MATHS

1. Which one of the following is the value of y when a = 5 and b = 2?

$$y = (a + 3)(4.5 - b)$$

A 72   C 20
B 54   D 18

2. Which one of the following is the value of y for the equation:

$$\frac{y - 2}{5} = y - 4$$

A 5.0   C 3.0
B 4.5   D 2.5

3. Which one of the following is the value of:

$$\frac{10^{2} \times 10^{-8}}{10^{-3} \times 10^{-6}}$$

A 10^3 C 10^1
B 10^2 D 10^0

4. Which one of the following is the value of c and d for the simultaneous equation?

$$c - d = 10 \qquad [1]$$
$$c + d = 50 \qquad [2]$$

A c = 30, d = 20   C c = 60, d = 30
B c = 40, d = 60   D c = 20, d = 40

5. Which one of the following transposes the equation to find I from:

$$S = \frac{\sqrt{I^2 t}}{k}$$

A $I = S^2k^2/t$ C $I = S^2k/t^2$
B $I = \sqrt{Sk^2/t}$ D $I = \sqrt{S^2k^2/t}$

6. The earth fault loop impedance Z_S for a consumer's circuit is given by the equation $Z_S = Z_E + R_1 + R_2$. Which one of the following gives the equation for finding R_2?

A $R_2 = Z_S + Z_E - R_1$ C $R_2 = Z_S - Z_E - R_1$
B $R_2 = Z_S + Z_E + R_1$ D $R_2 = Z_S - Z_E + R_1$

7. Using your calculator, which one is the solution to the problem, rounded off to two significant figures?

$$\frac{\sqrt{1.9} \times 7.5 \times 10 \times 100}{2.2 \times 20 \times 5}$$

A 120   C 65
B 83   D 47

8. Using your calculator, which one is the solution to the problem to three decimal places?

$$\frac{\sqrt{3} \times 415 \times 18 \times 0.7}{0.85}$$

A 55365.882   C 6950.454
B 10655.169   D 2050.588

9. Using your calculator, which one is the solution to the problem to two decimal places?

$$\frac{1}{\sqrt{7.517}} + \frac{1}{3.625^2}$$

A 0.64   C 0.44
B 0.52   D 0.41

10. What is the value of x?

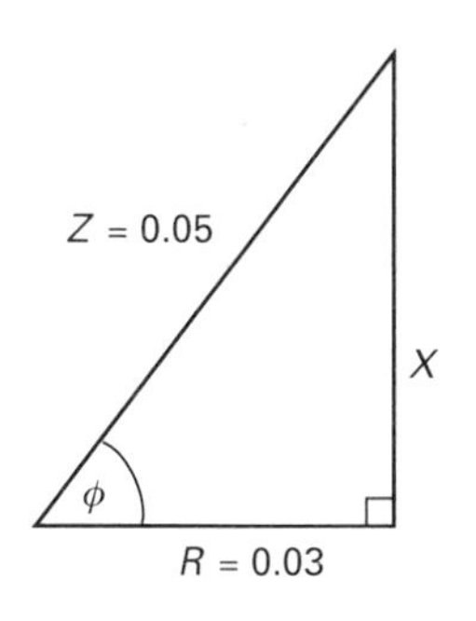

A 0.04 C 0.02
B 0.03 D 0.01

11. With reference to the above figure the angle phi (ϕ) is:

A 53.1° C 36.9°
B 45.0° D 19.7°

12. If the radius of a circle is 15 cm, its circumference in metres is:

A 1.00 m C 0.47 m
B 0.94 m D 0.30 m

13. Which numerical ratio is equal to cos 45°?

A $3/\sqrt{2}$ C $1/\sqrt{2}$
B $2/\sqrt{2}$ D $0/\sqrt{2}$

14. In a right-angled triangle, the ratio adjacent/hypotenuse is called:

A tan ϕ C sec ϕ
B sin ϕ D cos ϕ

15. What is the length of the hypotenuse side in the right-angle triangle with two sides equal?

A 3 C $\sqrt{3}$
B 2 D $\sqrt{2}$

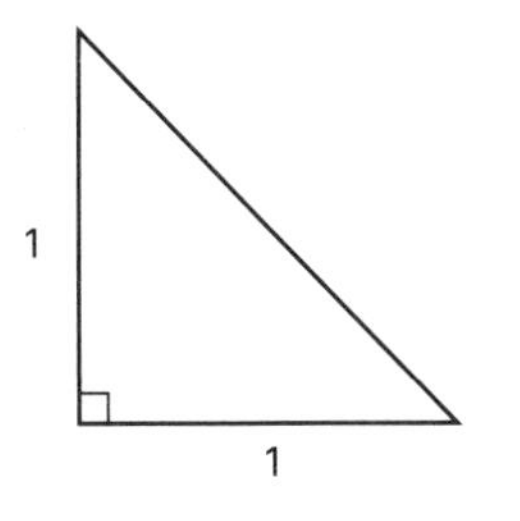

16. Which one of the following is equal zero?

A sin 90° C tan 90°
B cos 90° D sin 45°

17. The ratio $\sqrt{3}/2$ is equal to:

A sin 60° C sin 30°
B sin 45° D sin 15°

18. The length of BC is:

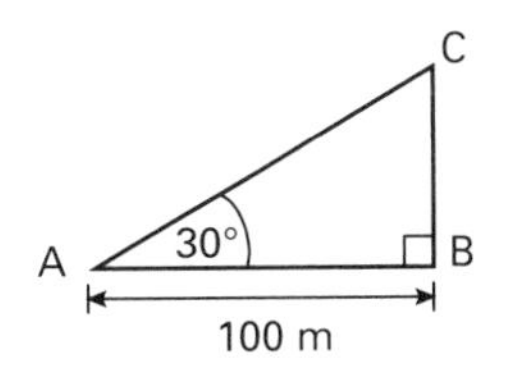

A 100.0 m C 57.74 m
B 86.75 m D 50.00 m

19. What is the length of side b?

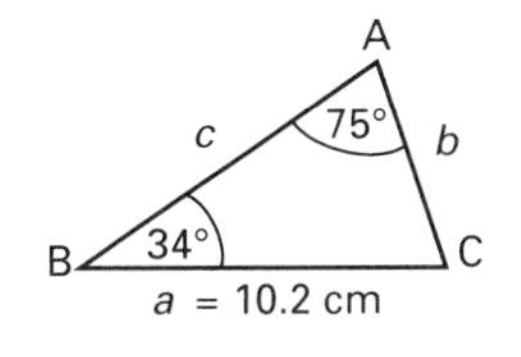

A 9.90 C 5.90
B 6.90 D 3.90

20. What is the length of side c in the above figure?

A 9.985 C 8.996
B 9.652 D 8.548

ALTERNATING CURRENT CIRCUITS

21. In Figure 2.1 the line indicated by the value 0.707 touches the sine wave in the positive direction at:

A 50° and 140° C 45° and 135°
B 45° and 130° D 50° and 135°

22. The unit of all the following circuit quantities is the ohm EXCEPT:

A resistance C reactance
B capacitance D impedance

23. A component with a zero power factor condition consumes:

A 100% power C unity power
B 50% power D no power

24. An inductor is a component possessing:

A resistance and inductance
B resistance and capacitance
C capacitance and inductance
D resistance only

25. Increasing the supply frequency to an inductor makes its reactance:

A smaller C larger
B unchanged D oscillate

26. The total power taken by a 3-phase, 415 V balanced load handling a supply current of 2 A with a power factor of 0.8 lagging is:

A 1992 W C 831 W
B 1150 W D 664 W

27. A 5 kW motor has a power factor of 0.7 lagging. The effect on the circuit with a 6 kVAr capacitor bank connected across its terminals is to create a condition of:

A lagging power factor
B leading power factor
C unity power factor
D zero power factor

28. The current taken by a 6 kW/240 V induction motor having a power factor of 0.75 lagging is:

A 33.33 A C 18.75 A
B 25.00 A D 11.11 A

29. The power factor of a purely resistive component connected to an a.c. supply is:

A unity C leading
B lagging D zero

30. If the line current taken by a delta connected load is 20 A, the phase current is:

A 11.55 A C 6.66 A
B 10.00 A D 3.33 A

31. An a.c. connected load at unity power factor takes a current of 12.5 A. What current does it take at 0.5 power factor?

A 50.00 A C 6.25 A
B 25.00 A D 0.00 A

32. Which type of component on a.c. provides a leading power factor?

A motor windings
B fire-bar element
C capacitor
D discharge lamp ballast

33. The value of Q is:

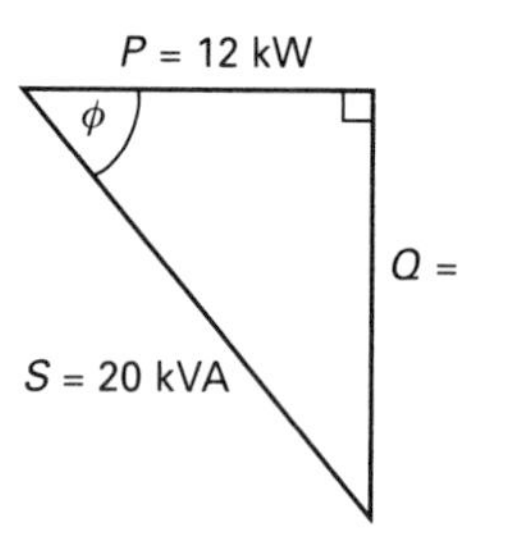

A 28 kVAr C 18 kVAr
B 20 kVAr D 16 kVAr

34. The power factor in the above figure is:

A 0.90 C 0.75
B 0.85 D 0.60

35. An a.c. circuit most likely to produce this phasor diagram is:

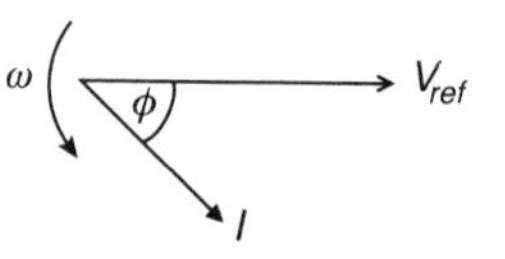

A resistance in series with capacitance
B inductance in series with resistance
C resistance in parallel with capacitance
D capacitance in parallel with inductance

36. The current taken by the circuit is:

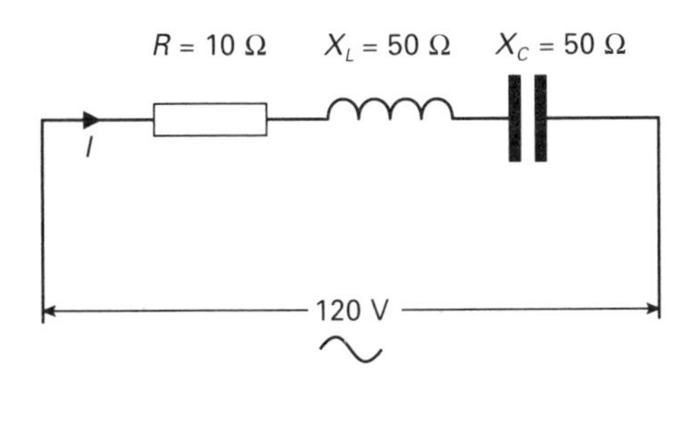

A 12 A
B 10 A
C 3 A
D 2 A

37. The power factor in the circuit above is:

A unity
B leading
C lagging
D zero

38. If the frequency of supply in the circuit for question 36 were 50 Hz the inductance of the coil would be:

A 15.90 H
B 1.59 H
C 159.0 mH
D 15.9 mH

39. The potential difference across the capacitor in the above figure is:

A 600 V
B 500 V
C 400 V
D 100 V

40. The particular frequency which causes X_L to equal X_C is called:

A ultra high frequency
B very high frequency
C medium frequency
D resonant frequency

ELECTRIC MOTORS AND STARTERS

41. A straight conductor of length 0.1 m is placed in a magnetic field of magnetic flux density 20 T. When a current of 50 mA is passed through it, it exerts a force of:

A 100 N
B 10 N
C 1 N
D 0.1 N

42. A motor develops an output power of 5 kW when running at 24 rev/s. The torque it produces is:

A 42.53 Nm
B 33.16 Nm
C 21.98 Nm
D 15.31 Nm

43. The efficiency of a motor having a rated output of 10.5 kW and total losses of 800 W is:

A 92%
B 80%
C 75%
D 61%

44. The rotating magnetic field inside the stator of a three-phase induction motor travels at:

A asynchronous speed
B slip speed
C synchronous speed
D rotor speed

45. A centrifugal switch is used in a single-phase motor to:

A open or close an auxiliary winding
B protect it from unintentional starting
C short circuit rotor bar conductors
D introduce reactance in the running winding

46. An a.c. motor having a synchronous speed of 25 rev/s and rotor speed of 24 rev/s produces a per unit slip of:

A 0.05
B 0.04
C 0.03
D 0.02

47. Which one of the following motors requires a wound rotor?

A Universal series motor
B Capacitor-start motor
C Shaded-pole motor
D Slip-ring motor

48. To reverse the direction of rotation of a three-phase cage induction motor, you should:

A create a delta connection for the stator windings
B create a star point at one end of the stator windings
C change the connections of two supply phases
D change the connections of all three stator windings

49. In the control circuit of a direct-on-line contactor starter, stop buttons are connected in:

A parallel with the no-volt coil
B series with the no-volt coil
C parallel with the overload trip
D series with the main windings

50. A reason for giving a motor starter no-volt protection is to avoid the motor:

A restarting after a supply failure
B taking excess current in its windings
C running at different speeds
D running at excessive temperature

51. Figure 3.10 shows The circuit diagram is of a three-phase:

A star-delta-start, cage rotor induction motor
B direct-on-line-start, cage rotor induction motor
C autotransformer-start, cage rotor induction motor
D resistance-start wound rotor induction motor

52. Except where stated in a relevant British Standard, the IEE Wiring Regulations require every motor exceeding 0.37 kW to be provided with control equipment incorporating a means of:

A surge control
B speed control
C reversing direction
D overload protection

53. To allow a motor to be stopped and started remote from its working:

A start buttons are wired in series and stop buttons wired in parallel
B start buttons and stop buttons are all wired in series
C start buttons are wired in parallel and stop buttons wired in series
D start and stop buttons are all wired in parallel

54. Which one of the following diagrams is correct for the rotor direction of an induction motor?

A (1) C (3)
B (2) D (4)

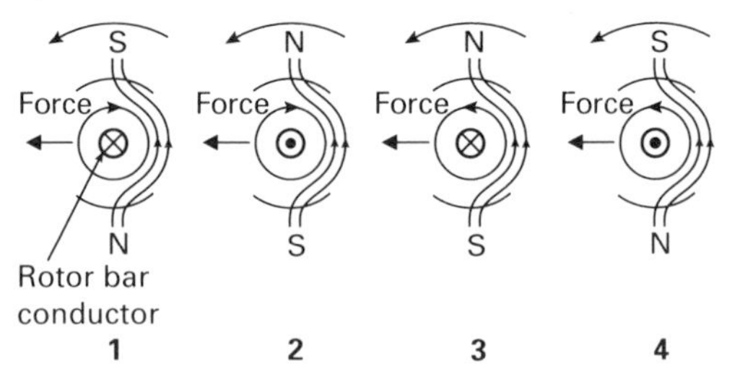

55. When a 415 V a.c. motor is connected to a star-delta starter, its star windings receive:

A 87% of the supply voltage
B 58% of the supply voltage
C 45% of the supply voltage
D 10% of the supply voltage

56. A three-phase cage induction motor has capacitors connected across its windings in order to improve the circuit:

A power losses C voltage drop
B power factor D protection

57. Which one of the following circuits shows the correct connections for a split-phase resistance-start induction motor?

A (a) C (c)
B (b) D (d)

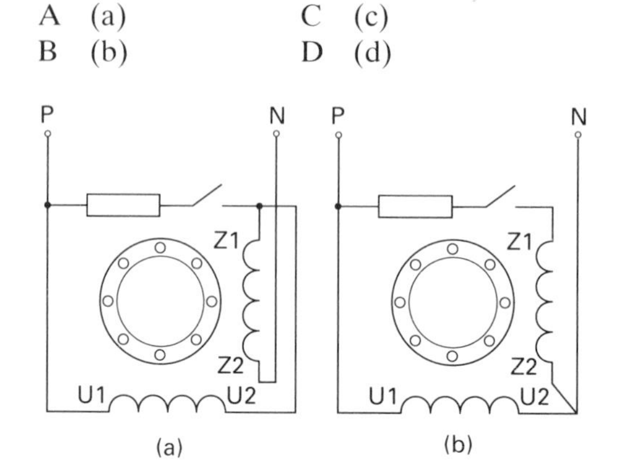

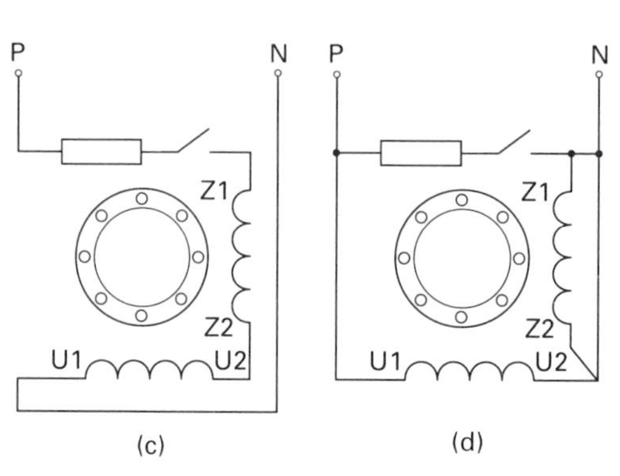

58. The turning effort of a motor's shaft is called:

A torque C force
B slip D acceleration

59. Which one of the following would protect a motor winding from overheating?

A residual current device
B thermistor
C no-volt coil
D fuse

60. The winding end connections of a three-phase induction motor are:

A X, Y, Z C R, Y, B
B M, N, O D S, T, P

ELEMENTS OF LIGHTING DESIGN

61. Which one of the following is not a colour of the visible light spectrum?

A yellow C green
B brown D red

62. A GLS incandescent lamps mostly emits:

A ultra-violet energy
B infra-red energy
C radio wave energy
D gamma-ray energy

63. The colour of a light source which reveals the colour of an object is called colour-

A matching C rendering
B appearance D visibility

64. Which one of the following is used in the determination of utilisation factor?

A room index
B maintenance factor
C working plane
D ceiling cavity

65. The ratio lumens/watt is called:

A efficiency C efficacy
B luminous flux D illuminance

66. A lighting luminaire has controlgear losses of 10 W and twin lamps rated at 70 W each. If each lamp produces 5000 lumens, the efficacy of the whole luminaire is:

A 71.43 lm/W C 58.84 lm/W
B 66.67 lm/W D 35.97 lm/W

67. Which one of the following is called a SON lamp?

A High-pressure sodium vapour discharge lamp
B Low-pressure sodium vapour discharge lamp
C High-pressure mercury vapour discharge lamp
D Low-pressure mercury vapour discharge lamp

68. A fluorescent tube fails to strike and there is no end glow from the tube. All the following are possible causes EXCEPT:

A faulty tube
B open-circuit choke
C faulty starter
D low voltage

69. A low-pressure SOX lamp produces a poor light output which appears mostly red. All the following are possible causes EXCEPT:

A failing lamp
B wrong lamp ballast
C lamp voltage too high
D lamp voltage too low

70. All the following lamps have excellent colour rendering qualities EXCEPT:

A GLS lamp C MBF lamp
B SON lamp D MBI lamp

Appendix 2

Written questions

1. Figure A2.1 shows a block diagram, of some important components in the control of a washing machine. Briefly describe the operation of the components marked A, B, C, D, E and F.

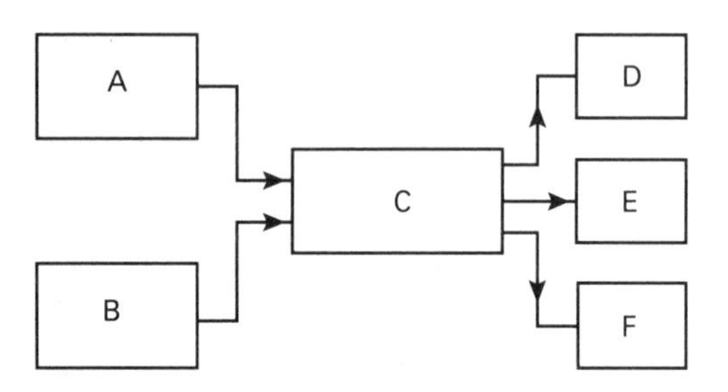

Legend A – water level sensor
B – water temperature sensor
C – controller
D – pump
E – heater
F – drum motor

Figure A2.1 Washing machine control

2. Figure A2.2 shows a block diagram of some important components in the control of a space heating system. Briefly describe the operation of the components marked A, B, C, D, E, F, and G.

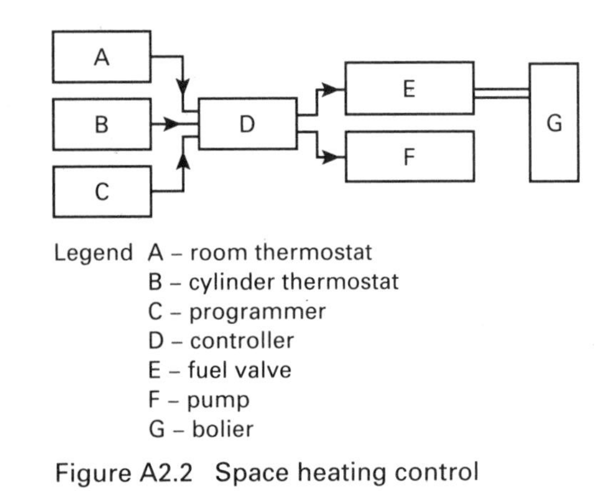

Legend A – room thermostat
B – cylinder thermostat
C – programmer
D – controller
E – fuel valve
F – pump
G – bolier

Figure A2.2 Space heating control

3. Figure A2.3 shows a block diagram of some of the important components in the speed control of a motor. Briefly describe the operation of the components marked A, B, C, D and E.

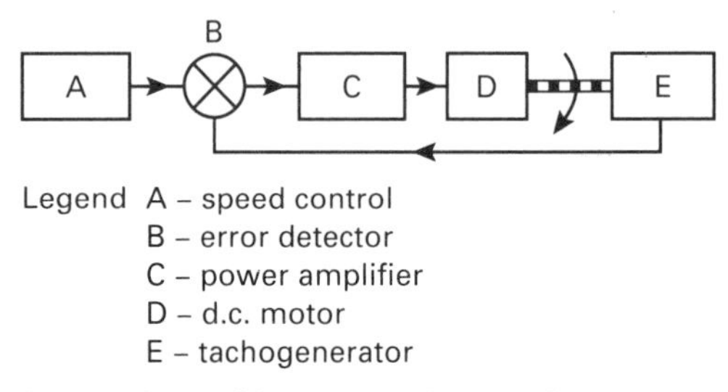

Legend A – speed control
B – error detector
C – power amplifier
D – d.c. motor
E – tachogenerator

Figure A2.3 Motor speed control

4. Figure A2.4 shows a block diagram of some important components in the control of a d.c. power supply. Briefly describe the operation of the components marked A, B, C and D.

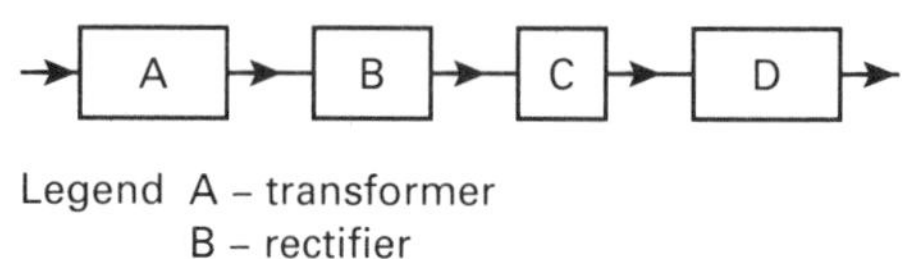

Legend A – transformer
B – rectifier
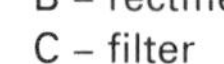
C – filter
D – stabiliser

Figure A2.4 Power supply

5. Figure A2.5 shows a block diagram of some important components in the control of a lamp dimmer. Briefly describe the operation of the components marked A, B, C, D, E, F and G

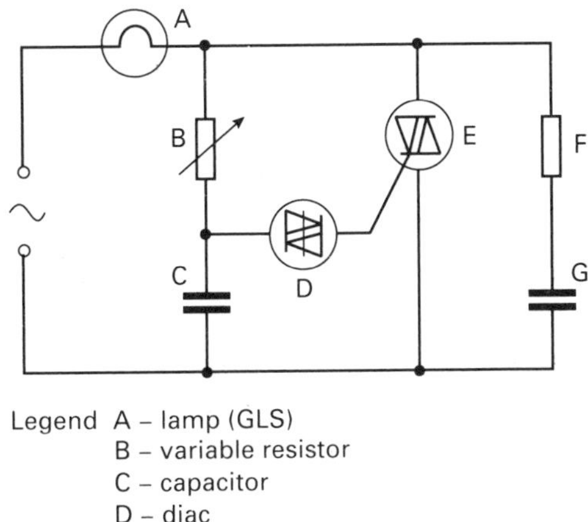

Legend A – lamp (GLS)
B – variable resistor
C – capacitor
D – diac
E – triac
F – resistor
G – capacitor

Figure A2.5 Lamp dimmer

6. Figure A2.6 shows a block diagram of some important components in the control of a security system. Briefly describe the operation of the components marked A, B, C and D.

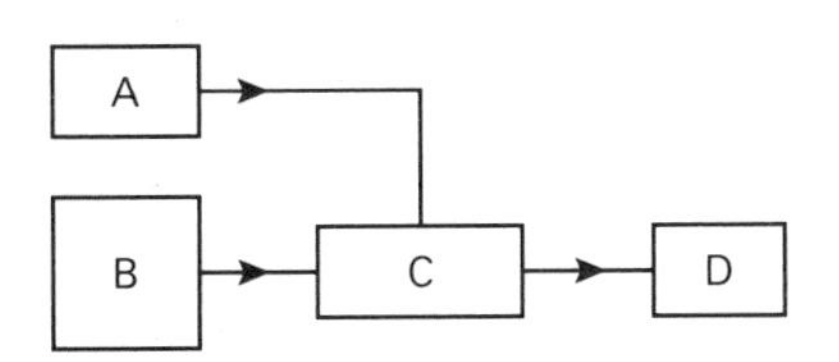

Legend A – personal attack button
B – sensing devices
C – control unit
D – alarm

Figure A2.6 Security system

7. (a) Draw a neatly labelled circuit diagram of a single-phase capacitor-start induction motor connected to the a.c. supply.
(b) State how reversal of direction is achieved.
(c) Describe the device inside a push button type starter which can operate in the event of a supply failure.

8. (a) Draw the circuit diagram of a low-pressure sodium vapour discharge lamp connected to a single-phase supply.
(b) Describe how the discharge is started in the SOX lamp.
(c) Explain why the SOX lamp is only suitable for certain applications. State TWO of its uses.

9. (a) What is meant by the term 'power factor'?
(b) The three-phase loads in a factory are as follows:
(i) 120 kW of heating at unity power factor
(ii) 240 kVA of inductive load at 0.8 power factor lagging.
Determine the total kVA and overall power factor of the two loads assuming they are balanced. Draw a phasor diagram.

10. (a) Explain the terms:
(i) colour rendering and
(ii) discomfort glare
(b) Figure A2.7 shows the lighting plan of a room drawn to a scale of 1:100. Each luminaire is fitted with a discharge lamp rated at 40 W/240V and having an efficacy of 100 lm/watt. If the utilisation factor and maintenance factors are 0.5 and 0.8 respectively, determine the average illuminance in the room.

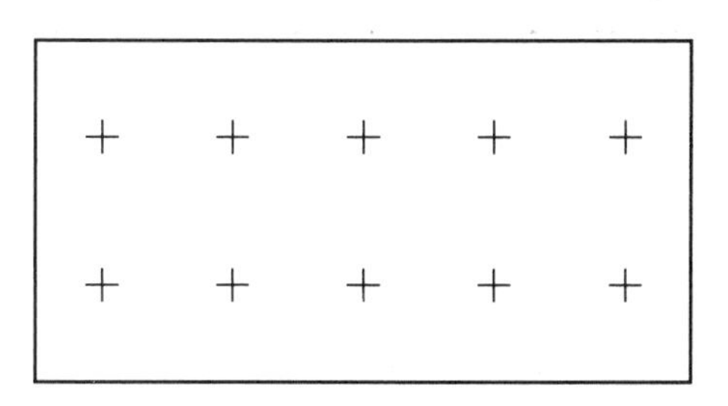

Figure A2.7 Lighting layout

Appendix 3

Answers

EXERCISE 1.1

1. $ab(a^2 + b)$

2. a^4b^3

3. $ABC = ABN + ACN$

$= \tfrac{1}{2}BNh + \tfrac{1}{2}CNh$

$= \tfrac{1}{2}h(BN + CN)$

but since

$BN + CN = BC$

$ABC = \tfrac{1}{2}BCh$

4. $A = lb - bh$, $A = 110\ \text{cm}^2$

5. $x = 5.5$

6. $a_1b_1 = a_2b_2$

but $b_2 = b_1 + 4$,

hence $a_1b_1 = a_2(b_1 + 4)$

$18b_1 = 16(b_1 + 4)$.

hence $b_1 = 32$ cm

$b_2 = 36$ cm

The common area is 576 cm^2

7. Let the inside area of the conduit be πr^2 and the outside area of the conduit be πR^2. Subtract the inside area from the outside area, thus:

$A = \pi R^2 - \pi r^2$

$= \pi(R^2 - r^2)$

$= \pi(R + r)(R - r)$

8. $V/I = V_1/I + V_2/I + V_3/I$

$= 1/I(V_1 + V_2 + V_3)$

The current cancels out on both sides of the equal sign, leaving:

$V = V_1 + V_2 + V_3$

9. $AB = x - 60$ A

$BC = x - 100$ A

$CD = x - 115$ A

$DE = x - 125$ A

$EF = x - 195$ A

$FG = x - 240$ A

10. $X = (S^2 - P^2)^{1/2}$

or $X = \sqrt{S^2 - P^2}$

EXERCISE 1.2

1. $y = 3$

2. $d = 1/5$

3. $n = -15$

4. $a = 6$

5. $x = -105$

6. 9.106×10^{-4}

7. 3.269×10^{-2}

8. 3.24×10^5

9. 145.67 or 1.46×10^2

10. 9.21×10^3

11. $x = 2,\ y = -1$

12. $x = -1,\ y = 6$

13. $m = 3,\ n = -5$

14. $p = 3,\ q = 1$

15. $I_1 = 2$ A, $I_2 = -1$ A

16. AC = 52.82, CB = 113.288

17. $R = 22.5\ \Omega$, $X = 19.84\ \Omega$

18. $P = 146.97$ kW, $\phi = 45.58°$

19. $\phi = 52.18°$

20. $S_1 = 41.54$ kVA, $Q_1 = 33.175$ kVAr

$S_2 = 27.37$ kVA

$Q = Q_1 - Q_2$

$= 33.175 - 11.13$

$= 22.04$ kVAr

EXERCISE 2

1. See page 116 of '*Questions and Answers in Electrical Installation Technology*' by same author.

2. a) (i) $X_L = 62.84\ \Omega$

(ii) $Z = 65.95\ \Omega$

(iii) $I = 3.64$ A

(iv) $V_R = 72.8$ V, $V_L = 228.7$ V

(v) $\cos\phi = 0.3$ lagging

(vi) $P = 265$ W

b) see Figure A3.1

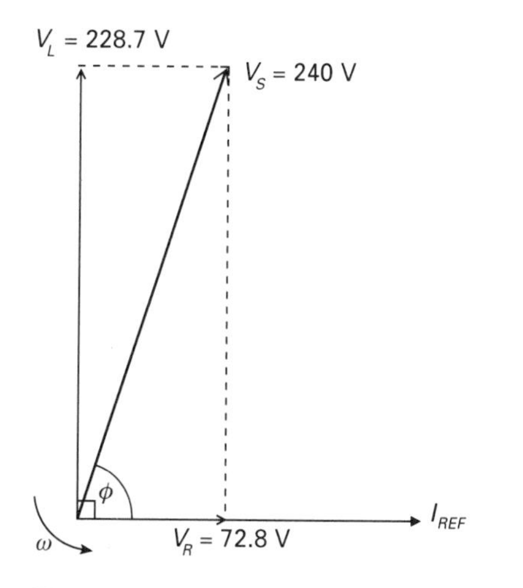

Figure A3.1

3. a) (i) $X_C = 159.1\ \Omega$

(ii) $Z = 166.8\ \Omega$

(iii) $I = 1.44$ A

(iv) $V_R = 71.9$ V, $V_C = 229$ V

(v) $\cos\phi = 0.3$ leading

(vi) $P = 103.68$ W

b) see Figure A3.2

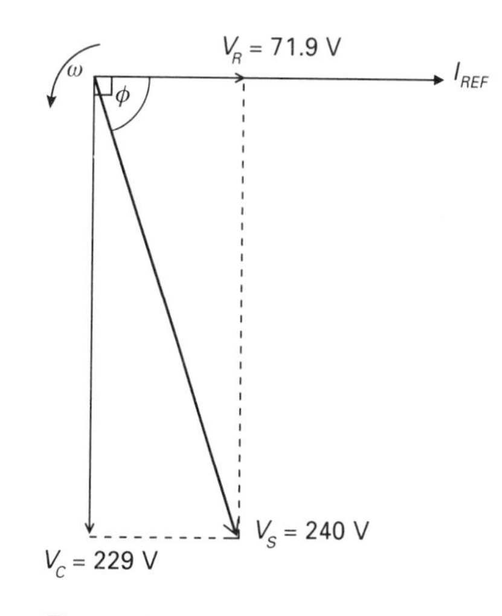

Figure A3.2

4. a) (i) $Z = 50\ \Omega$

(ii) $I = 3$A

(iii) $V_R = 90$ V, $V_L = 150$ V, $V_C = 270$ V

(iv) $\cos\phi = 0.6$ leading

(v) $P = 270$ W

b) see Figure A3.3

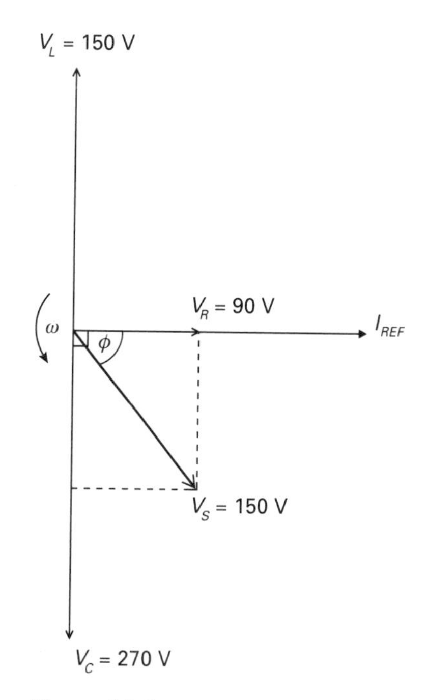

Figure A3.3

5. See page 117 of '*Questions and Answers in EIT*' by same author.

6. See page 98 of '*Questions and Answers in EIT*' by same author.

7. Your graph should be constructed along the lines of Figure 2.2. Use a scale of 1 cm = 1 A and 1 cm = 100 V. Answer: I = 2.12 A, V = 212 V

8. See page 119 of '*Questions and Answers in EIT*' by same author

9. See page 93 of '*Questions and Answers in EIT*' by same author

10. See page 90 of '*Questions and Answers in EIT*' by same author.

EXERCISE 3

1. See Figure A3.4

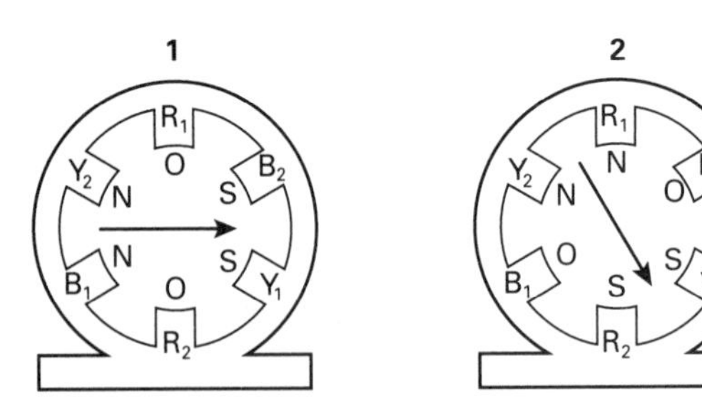

Figure A3.4 Phase positions

2. a)

	n_s (rev/s)	n_r (rev/s)
(i)	50	48.5
(ii)	25	24.25
(iii)	20	19

b) The term is used to describe a fault condition resulting from an open circuit in a winding of a three-phase motor. It is often caused by overheating and shorting out of the winding, leading to the rupturing of line fuses.

3. a) See Figure A3.5

b) Anticlockwise

c) Reverse the connections of its starting winding.

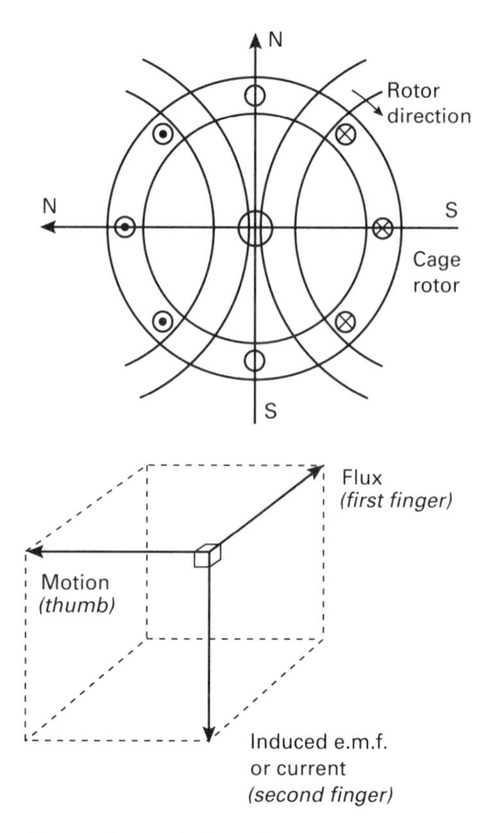

Figure A3.5 Induced currents in rotor bar conductors as a result of travelling magnetic field (found by applying Fleming's right-hand rule). Note the polarity of rotor.

4. a) Its main advantage is that it provides a much higher starting torque. A disadvantage is in the extra cost of equipment (more expensive motor and need for external rotor resistance).

b) Start buttons are connected in parallel while stop buttons are connected in series.

c) It mainly serves two purposes, namely:

(i) acts as the main contactor coil and

(ii) stops the motor from re-starting automatically, immediately after a supply failure.

5. (a) See Figure A3.6

(b) P = 1.33 kW, I = 7.94 A,

n_r = 47.5 rev/s

Note: For further information, see pages 109/110 of 'Questions and Answers' book by same author.

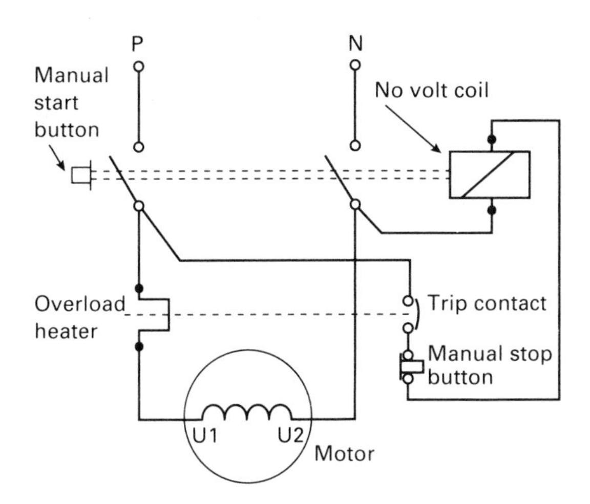

Figure A3.6 Single-phase motor connections

6. P_o = 13.9 kW

   η = 0.82 p.u.

   $p.f.$ = 0.62 lagging

   Note: For further information, see pages 107/108 of '*Questions and Answers*' book by same author.

7. See Figure 3.32.

8. See chapter notes.

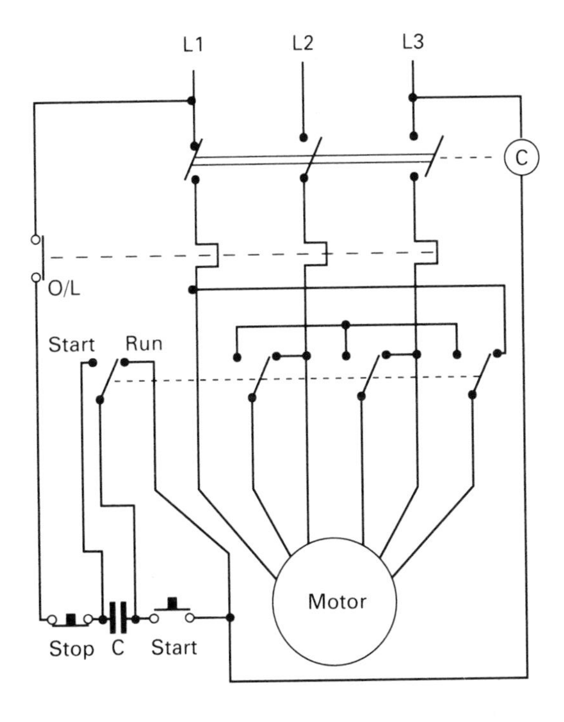

Figure A3.7 Hand-operated star-delta starter showing contactor control

9. The hoist is raised by pressing B1 which allows current to flow through the normally closed contact B2 and coil HR to complete the control circuit. This brings in contacts R1, R2 and R3 of the main circuit to operate the motor. To lower the hoist, B2 is pressed and current flows through the normally closed contact B1 and coil HL to complete the control circuit. This brings in contacts L1, L2 and L3 of the main circuit to operate the motor. A similar procedure operates the traverse motor. You should notice the main contact wiring to see how reversal of direction is achieved.

10. See Figure A3.7. The start button must be kept pressed until the switch is moved into the run position.

EXERCISE 4

1. a) See Figure A3.8
   b) See Chapter 4 text
   c) 9.75 lm/W

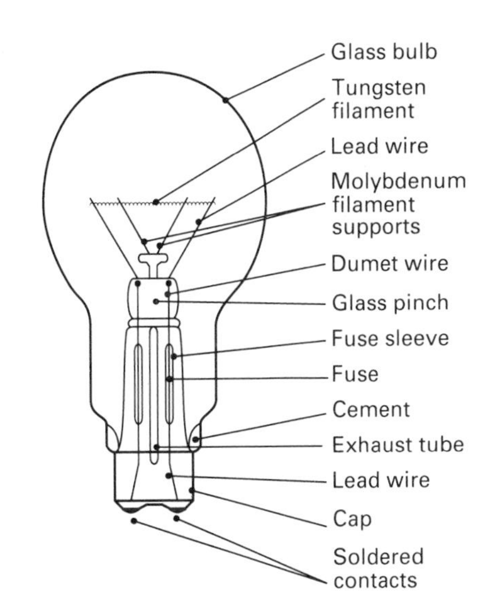

Figure A3.8 GLS lamps

2. a) See Chapter 3
   b) See Chapter 3
   c) See Figure A3.9

3. a) 1.58 m
   b) 100 lx

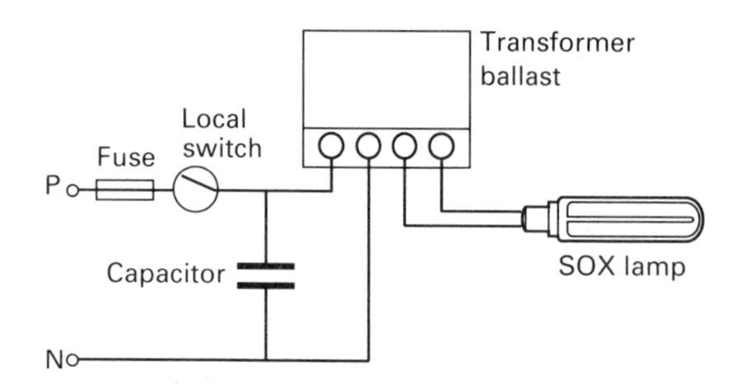

Figure A3.9 SOX lamp

4. a 10^5 lm; 25 luminaires

b) See Chapter 4 text

5. a) Enclose the luminaire with an opal plastic diffuser.

b) See Chapter 4 text

c) See Chapter 4 text

6. The operation and circuit diagram of this lamp are described on page 53 of the *Part 1 Studies Theory* book.

7. See page 56 of *Part 1 Studies Theory* book.

8. See page 57 of *Part 1 Studies Theory* book.

9. The answer to this question is explained in Appendix 2, Question 5.

10. Low supply voltage, perhaps volt drop.
Faulty lampholder contacts
End of lamp life
Capacitor for p.f. wrongly connected.

ANSWERS TO MULTIPLE-CHOICE QUESTIONS

1. C	**15.** D	**29.** A	**43.** A	**57.** B
2. B	**16.** B	**30.** A	**44.** C	**58.** A
3. A	**17.** A	**31.** B	**45.** A	**59.** B
4. A	**18.** C	**32.** C	**46.** B	**60.** A
5. D	**19.** C	**33.** D	**47.** D	**61.** B
6. C	**20.** A	**34.** D	**48.** C	**62.** B
7. D	**21.** C	**35.** B	**49.** B	**63.** C
8. B	**22.** B	**36.** A	**50.** A	**64.** A
9. C	**23.** D	**37.** A	**51.** D	**65.** C
10. A	**24.** A	**38.** C	**52.** D	**66.** B
11. A	**25.** C	**39.** A	**53.** C	**67.** A
12. B	**26.** B	**40.** D	**54.** D	**68.** D
13. C	**27.** B	**41.** D	**55.** B	**69.** C
14. D	**28.** A	**42.** B	**56.** B	**70.** B

ANSWERS TO WRITTEN QUESTIONS

1. A – The water level sensor is a pressure switch which switches off the supply to the heater element when sufficient water is obtained in the drum.

B – The water temperature sensor is a thermostat which switches off the washing machine heater elements when the desired temperature has been reached. It provides a range of temperatures between 0–87°C.

C – The controller is a rotary timing device incorporating electronic logic switching to sequence various washing operations such as fill, agitation, spin, rinse and drain, etc.

D – The drain pump is to remove water in the drum of the washing machine after each wash or rinse.

E – The heater is about 2.5 kW and is used for raising the water to the correct operating temperature.

F – The drum motor drives the drum at specified speeds given to it by the controller. The motor is often a universal motor designed to fast spin at speeds around or even higher than 16.67 rev/s.

2. A – The room thermostat operates on room temperature changes. Once set to a required level it will open or close the circuit according to the room temperature. If this is below the thermostat setting the circuit will be closed and the heating pump circulates hot water around the system.

B – The cylinder thermostat is the master control thermostat of the boiler and operates 'on' and 'off' at various selected temperatures of heated water. It is normally set above the room thermostat and when the correct temperature is reached it opens and closes a motorised flow valve which switches off the pump.

C – The programmer is the user's master control box for selecting desired hot water and heating times. It incorporates a time clock for automatic operation of the heating system every day of the week as well as several periods during each day. It also has override facilities of the pre-set times.

D – The controller is the device at the heart of the heating system, incorporating all the electrical connections to operate the boiler's fuel feed system, motorised valves, pump, cylinder thermostat, etc. The switching functions of the controller are often achieved electronically or through various relays and coils.

E – The fuel valve control is a magnetic on/off device for controlling the supply of fuel to the boiler's burner.

F – The pump is used for circulating water through the heating system's radiators. It is usually controlled by the room thermostat, although the cylinder thermostat needs to be closed to operate the boiler for heat.

G – The boiler provides the storage of the heated water. It incorporates the fuel burner and safety control valve should the burner fail to ignite or suffer a fault.

3. A – The set speed component supplies a reference signal to the error detector indicating the desired speed at which the d.c. motor should run.

B – The error detector receives the reference signal from the set speed component and also a feedback signal from the tachogenerator. It compares both signals and feeds the difference into the power amplifier.

C – The power amplifier, on receiving the error detector's signal, drives the motor at the required speed. A small feedback signal would indicate that the motor was running too slow, whereas a large signal would indicate that the motor was running too fast.

D – The d.c. motor is used to drive the mechanical load.

E – The tachogenerator is coupled to the driveshaft of the motor and the voltage it produces is fed back to the error detector. The faster the shaft turns the higher the feedback voltage will be.

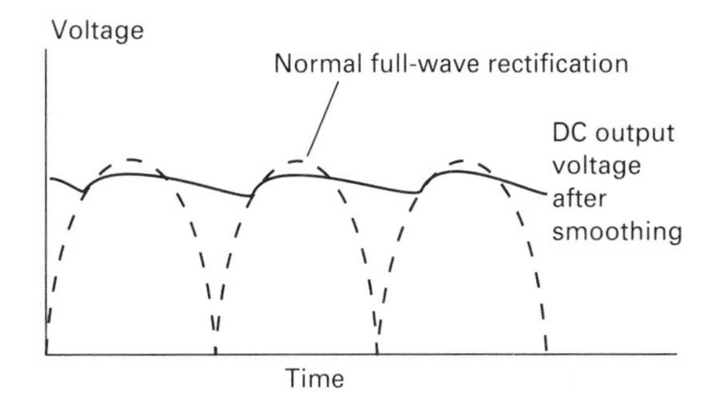

Figure A3.10 Results of using a filter in a power supply

4. A – The secondary winding of the transformer reduces the a.c. supply voltage to a lower value and feeds a full-wave bridge rectifier.

B – The rectifier changes the transformed voltage into full-wave d.c. using four semiconductor diodes. (For an explanation, see page 98 of *Electrical Installation Technology 2* by the same author.

C – The filter circuit smoothes out the unidirectional d.c. by eliminating the 'lumpy' full-wave of the rectified d.c. It contains a large capacitor called a reservoir capacitor and performs the function shown in Figure A3.10.

D – The stabiliser is a voltage regulator used to keep the d.c. output from the filter circuit at a fixed value and also reduce fluctuations due to load changes (see page 61–62 *Part 1 Studies: Science* book). Figure A3.11 is a typical power supply circuit diagram.

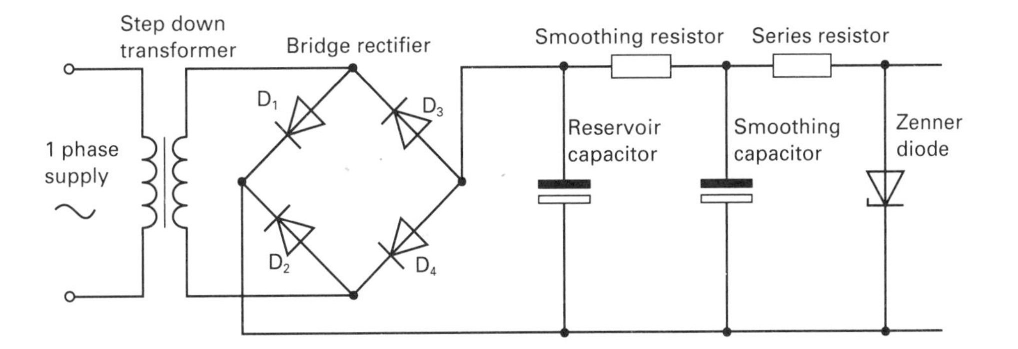

Figure A3.11 Power supply circuit

5. A – The GLS lamp has a tungsten filament which allows it to operate at about 2500°C. It is wired in series with the triac component.

B – The variable resistor is part of a trigger network providing a variable voltage into the gate circuit containing a series connected diac component. Increasing the value of the resistor, increases the time taken for the capacitor to reach its voltage level to pass current into the diac circuit. Reducing the resistance allows the triac to switch on faster in each half cycle. By this adjustment the light output of the lamp can be controlled from zero to full brightness.

C – The capacitor is connected in series with the variable resistor and both are designed to produce a variable phase shift into the gate circuit of the diac. When the p.d. across the capacitor rises, sufficient current flows into the diac to switch on the triac.

D – The diac is a triggering device having a relatively high switch-on voltage (35 V) and acts as an open switch until the capacitor p.d. reaches the required voltage level.

E – The triac is a two-directional thyristor which is triggered on both halves of each cycle. This allows it to conduct current in either direction of the a.c. supply. Its gate is in series with the diac, allowing it to receive positive and negative pulses.

F – A relatively high ohmic value resistor (100 Ω) in series with a capacitor to reduce false triggering of the triac caused by mains interference.

G – A relatively low value capacitor (0.1 μF) in series with F above and for the same reason. The RC combination is called a 'snubber circuit'.

6. A – The personal attack button is intended to activate the alarm system when pressed and is wired directly to the control panel. It is fixed in an accessible position, near a front door or adjacent to a bed.

B – The sensing devices are types of detector used to initiate the alarm if an intruder is present. They may be a combination of magnetic switches, pressure matts or passive infra-red detectors and wired in series or parallel back to the control panel.

C – The control unit, in its simplest form, is often key-operated, incorporates a touch sensitive key pad, zone display, PA indicator, exit/entry and tamper facilities. Inside the unit will be a battery pack and the wiring connections to all the detectors, visual strobe light and alarm bell.

D – The alarm bell is often housed in a box and placed in an inaccessible position high up on an outside wall. It will incorporate a tamper switch to stop the box cover from being removed.

For further information of this circuit, see pages 63–65, *Part 1 Studies Theory* book by the same author.

7. a) See Figure 3.23

b) See Chapter 3 regarding single-phase induction motors.

c) This is called the 'under-voltage' or 'no-volt' release coil and is explained in Chapter 3 under the sub-heading of motor protection.

8. See page 57 of *Part 1 Studies Theory* book by the same author.

9. (a) See 'power factor' in chapter 2 of this book.

(b) Answer 343.6 kVA. See Q152 of *Questions and Answers in Electrical Installation Technology* by the same author. Figure A3.12 is a phasor diagram of the circuit.

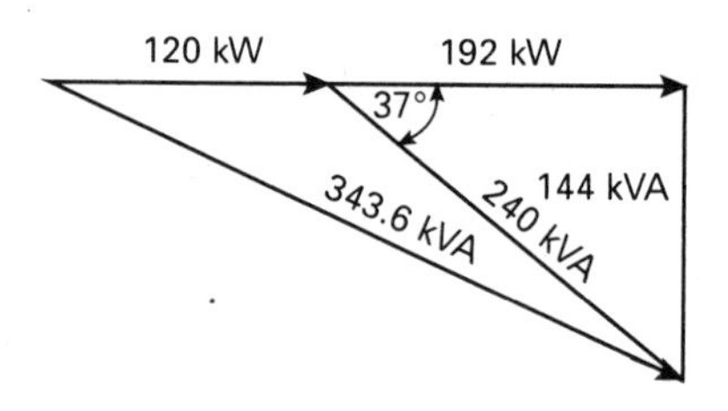

Figure A3.12 Phasor diagram

10. (a) See chapter 4 of this book.

(b) E = 400 lx.